TOBRADEX ST
(tobramycin 0.3% and dexamethasone 0.05%
Ophthalmic Suspension)

D1088785

**THE BATTLESHIP
BISMARCK**

ULRICH ELFRATH   BODO HERZOG

THE BATTLESHIP
BISMARCK

A Documentary in Words and Pictures

ULRICH ELFRATH   BODO HERZOG

Schiffer Publishing Ltd

1469 Morstein Road, West Chester, Pennsylvania 19380

Translated from the German by Dr. Edward Force, Central
Connecticut State University.

Printed in the United States of America.
ISBN: 0-88740-221-6

This title originally published under the title, "Schlachtschiff
Bismark-Ein Bericht in Bildern und Dokumentation," by Podzun-
Pallas-Verlag, Friedberg 3 (Dorheim/H), © 1975, ISBN:
3-7909-0029-X.

Published by Schiffer Publishing, Ltd.
1469 Morstein Road
West Chester, Pennsylvania 19380
Please write for a free catalog.
This book may be purchased from the publisher.
Please include $2.00 postage.
Try your bookstore first.

Contents

Foreword

The 277 days of the battleship BISMARCK—from its launching to its sinking—form one of the most gripping dramas to take place at sea during World War II.

During that short span of service time, it was the largest battleship in the world. The superior technology on board, the mighty weapons and the high level of the crew's training made this ship a seemingly invincible war machine. In the first light of morning on May 25, 1941, the battleship scored its triumphal success over the mighty British battle cruiser HOOD—England's pride.

The storm of joy at the victory was followed by hours of hope of trying to escape the far-flung net of the Royal Navy, and the moment of recognition that the safety of a harbor on the Atlantic coast of France could no longer be attained.

Then the encircling of the BISMARCK was complete. The heavy British units had surrounded the German battleship. The fight to the death began.

It was indeed a dramatically overwhelming fate for the men and the ship. Many books have been written about the battleship BISMARCK and the events of its victory and defeat. One of the best among them is surely the book by J. Breenecke, *SCHLACHTSCHIFF BISMARCK.*

A thoroughgoing photo documentation with a richness of photos of the ship, its equipment, weaponry, armor, fire control and much more, was previously lacking. The publishers are indebted to the authors, Mr. Ulrich Elfrath and Mr. Bodo Herzog, for the intensity with which they went about the difficult task of finding, authenticating and arranging pictures.

Only through the personal involvement of the authors was it possible to attain this wealth of photos, of detail shots of the ship, and of pictures taken in battle. For those who can absorb the optical impressions of the course of events, this book will be of high value, as well as for those enthusiasts who are model builders, or "ship lovers" of the great vessels.

The Launching

Wooden Model of the Bow Section of the BISMARCK

From the start, military thinking in Germany was influenced in terms of the continent, for politically expedient strivings to expand were limited to Europe itself, quite unlike England, which because of its geographical situation had to develop a maritime consciousness if it wanted to survive. The basis of its offensive and defensive forces was a strong fleet. In Germany, on the other hand, a naval fighting force played—if any—only a very subordinate role. A national navy existed only since 1871. In comparison with the land forces, which were favored ideally and materially and, measured by international standards, soon ranked among the strongest, the fleet was built up only hesitantly, just as before. Only a few years before World War I, these delays were supposed to be made up for in a "generation jump." In a mighty effort, Imperial Germany built up a naval fleet second only to that of Great Britain in the world. During World War I, when the Royal Navy exceeded the tactical range of the German capital ships, it was obvious that the result of the national sacrifice that had been invested in expanding the fleet, was the greatest military failure up to that time.

After World War I the strategic sea position could be improved only slightly by technical (such as increased ranges) and tactical (such as supplying at sea) measures. The use of major warships also had to remain questionable because military aircraft, as far as could be foreseen, would become a serious threat to them. Still in all, the German Navy wanted to be represented again by major warships.

The battleship BISMARCK was the first new ship to be built as part of a fleet program that would culminate in the building of super-battleships of over 110,000 tons.

The launching took place on Tuesday, February 14, 1939. Amid the mighty construction facilities of the Blohm & Voss shipyards in Hamburg, in the shadow of the BISMARCK's gigantic hull, Adolf Hitler delivered the launching address. The battleship was subordinated to the goals of the National Socialist regime.

According to the concepts then in effect, it took on an aggressive tone: ". . . Princes and dynasties, political fence-sitters and Social Democrats, liberalism, state parliaments and Reichstag parties no longer exist. All of them, that had once caused such difficulties for the historical strivings of this man (i.e., Bismarck), had outlived him by only a few decades. But National Socialism, in its development and in the German community of peoples, had created the elements of the spiritual world-view and the organization that were capable of defeating the enemies of the Reich now and forever . . ." (Source: "Völkischer Beobachter", No. 46, Feb. 15, 1939).

Adolf Hitler closed his speech before the heads of the National Socialist regime with the emotional words:

"German builders, engineers and workers have created the mighty hull of this proud giant of the sea. May the German soldiers and officers who have the honor of utilizing this ship always prove to be worthy of its namesake! May the spirit of the Iron Chancellor be transmitted to them, may it accompany them in all their actions on their fortunate journeys in peace, but if it should ever be necessary, may it lead them and be remembered by them in the hour of the hardest fulfillment of their duty!

"With this fervent wish, the German people welcome their new battleship BISMARCK."

The christening was done by Bismarck's granddaughter Dorothea von Loewenfeld—the name board and arms were unveiled, but the mighty hull of the BISMARCK did not move . . .

But then, at 1:34 P.M., it slowly glided into its new element! It was to be one of the strongest and most modern battleships of all the world's navies.

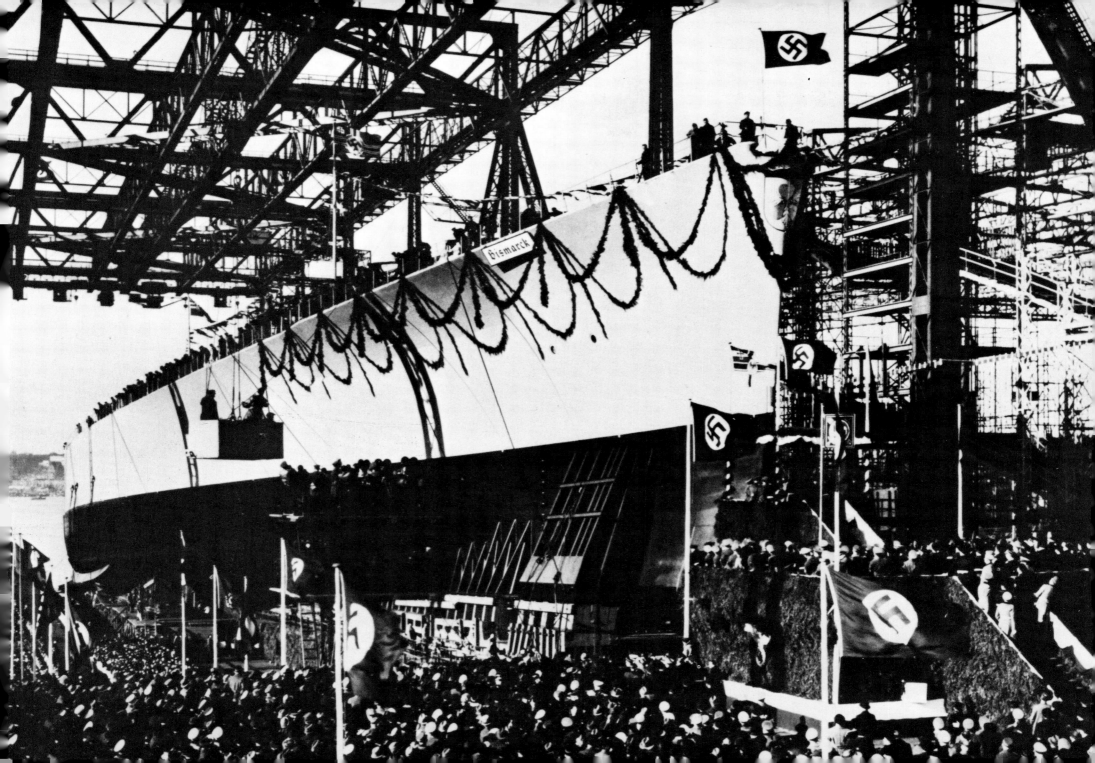

Technical Data, Fittings and Equipment

The BISMARCK was contracted for with the shipbuilding firm of Blohm & Voss under the project designation letter F. Laying the keel began on July 1, 1936. Some three years later, on February 14, 1939, the ship was formally launched. On the very next day, work was continued at the equipping pier, and earlier than planned, on August 24, 1940, the BISMARCK was put into service. Test cruises, at first in the North Sea and later in the Baltic, commenced immediately thereafter. In May of 1941 the battleship was declared ready for war service. (Kb)

Water displacement and dimensions

Construction displacement (calculated volume)	45,172 tons
Maximum displacement (including maximum load)	50,900 tons
Length at construction waterline (KWL) (based on construction displacement)	241.5 mts.
Overall length (LüA) (greatest length)	251.0 meters
Width (greatest width)	36.0 meters
Draught (at construction displacement)	8.7 meters
Draught (at maximum displacement)	10.2 meters

The weights are stated in English (long) tons (' 1.016 tons).

According to naval agreements in which the German and British navies had set up mutual limits in terms of complete and individual tonnage, the BISMARCK officially belonged in the 35,000-ton battleship class. With 44,700 tons (1920 construction level), the HOOD was then listed as the world's largest battleship. Even when, after her modernization in 1940, her displacement was increased to 48,500 tons, she was still smaller than the BISMARCK. The BISMARCK's victory over the HOOD was even upgraded at the time for propaganda use by the statement that the Royal Navy had lost the world's largest warship, though this status actually belonged to the BISMARCK. Only after the war did the actual size relationship become generally known.

According to the technical knowledge of the time, warship builders strove for a width-to-length ratio in favor of the length, in order to increase speeds. In the HOOD, this ratio, based on the construction waterline, was 1:8.2, while in the BISMARCK it was only 1:6.7. Despite this obviously better ratio for the HOOD than the BISMARCK, the HOOD, with almost the same displacement, had a speed advantage of only one knot, with simultaneous limitations on the ship's stability and seaworthiness.

In addition, the extra width of the BISMARCK allowed better space utilization, armor arrangement, and the housing of the medium artillery in twin turrets without omitting the ship's heavy anti-aircraft guns.

Machinery and Performance

Performance (3 propellers)	150,170 HP
Speed	30.1 knots
Fuel capacity (maximum)	8000 tons, oil-fired
Range	9280 nautical miles at 16 knots 8900 nautical miles at 17 knots

The BISMARCK's powerplant was planned for a top speed of 29 knots. Thus 46,000 horsepower were to drive each shaft, for a total of 138,000 HP. During testing it was shown that the specifications had been exceeded considerably. The turbines produced 150,170 HP for a speed of 30.1 knots. This made the ship one of the fastest battleships ever built.

The use of three shafts was necessary in the BISMARCK only because the battleships of other countries attained high speeds only with four propellers. But because of the central shaft, the BISMARCK had to have two parallel rudder systems (course stabilization) installed.

The BISMARCK, like all large units in the German Navy, was planned for long-range operations (oceanic naval warfare). For that reason, a large range was required, which was almost twice as great for the Bismarck as for comparable battleships of other countries; nevertheless, it turned out to be insufficient for the BISMARCK.

Machinery

The power was produced by three drive turbines for the three propeller shafts. These turbines had a high-, medium- and low-pressure section. At full load, the speeds for high and medium pressure were 2825 revolutions, for low pressure 2390 revolutions per minute. The turbines with their drive systems were housed in three turbine rooms; of them, the

Right page:
This picture expresses power and speed. At that time, battleships were still an expression of maritime power. Because of the lack of distance-measuring apparatus, the silhouette of the BISMARCK is not yet completed, but the "beauty of her architecture" is undoubtedly obvious.

September 1940:
The BISMARCK in Kiel Bay after her first passage through the Kaiser Wilhelm Canal.

powerplant for the middle propeller shaft was far aft; in front of it, almost amidships at the port and starboard sides, were the other two powerplants. The steam power was produced by twelve Wagner boilers. There was a boiler room for every four boilers and their accessories. The boiler rooms were divided by longitudinal bulkheads, so that in the end only two boilers were installed together.

Great quantities of energy were required to operate the ship. Just for the operation of the rotation and other operations of the ship's artillery, there were four electric plants, each with two 4000-kilowatt Diesel generators. In all, the electrical generators produced 7910 kilowatts at 220 volts.

Aircraft Equipment

The BISMARCK, similarly to the British battleships WARSPITE, MALAYA and RENOWN after their last rebuilding, was equipped with an airplane catapult running at a right angle to the ship's longitudinal axis. The catapult was located amidships between the funnel and the mainmast. According to the wind, airplanes could be catapulted off to both port and starboard (double catapult). The airplanes were kept ready in hangars, one of them in a readiness hangar beside the funnel and the other four in a hangar under the mainmast. Two cranes, also planned as boat cranes, lifted planes in the water back onto the catapult. From there they were returned to the hangars on rails.

Other Equipment

The battleship BISMARCK was an independent, self-sufficient battle unit. Naturally, this independence characterized the inner and outer architecture of the battleship.

The BISMARCK had to have "social spaces" with the appropriate supply, accommodation and hygienic facilities for more than 2000 servicemen.

The external appearance of the deck was influenced by the great number of motorboats: three captains' gigs (K-boats), four tenders (V-boats) and three longboats. There were also six cutters, which were used for outboard work and also for sporting events.

Less eye-catching but just as important for maintaining military operations were deck loudspeakers, signal lights, air shafts, cable and fire-hose reels, companion ways, ammunition racks in readiness for the anti-aircraft guns, etc., which completed the external picture of the BISMARCK.

Weights

The ship's construction units and equipment amounted to the following weights and proportions:

Ship's hull	12,700 tons	25.4%
Armor plate	17,256 tons	38.2%
Powerplant	2,756 tons	6.1%
Auxiliary equipment	1,400 tons	3/1%
Armament	7,453 tons	16.5%
Equipment	1,345 tons	3.0%
Fuel	3,388 tons	7.5%
Aircraft equipment	100 tons	0.2%
Construction displacement	45,172 tons	100%

Crew

The crew strength totaled 2092 men, of whom there were—
103 officers, including ensigns,
1962 non-commissioned officers and crewmen, and
27 men of all ranks for the prize crew.

The building costs of the BISMARCK were stated as 196.8 million Reichsmark. This statistic is not provable, but may be regarded fairly safely as probably being too low.

A boiler hangs on the crane lines . . .

. . . and is lowered into the ship's hull.

Assembly work on the propeller and rudder systems. The port rudder is already installed. The rudder blade has a surface area of 24 square meters; each of the screws has a diameter of 4.85 meters.

In one of the rudder machine rooms—pumps and steering machinery for the after steering.

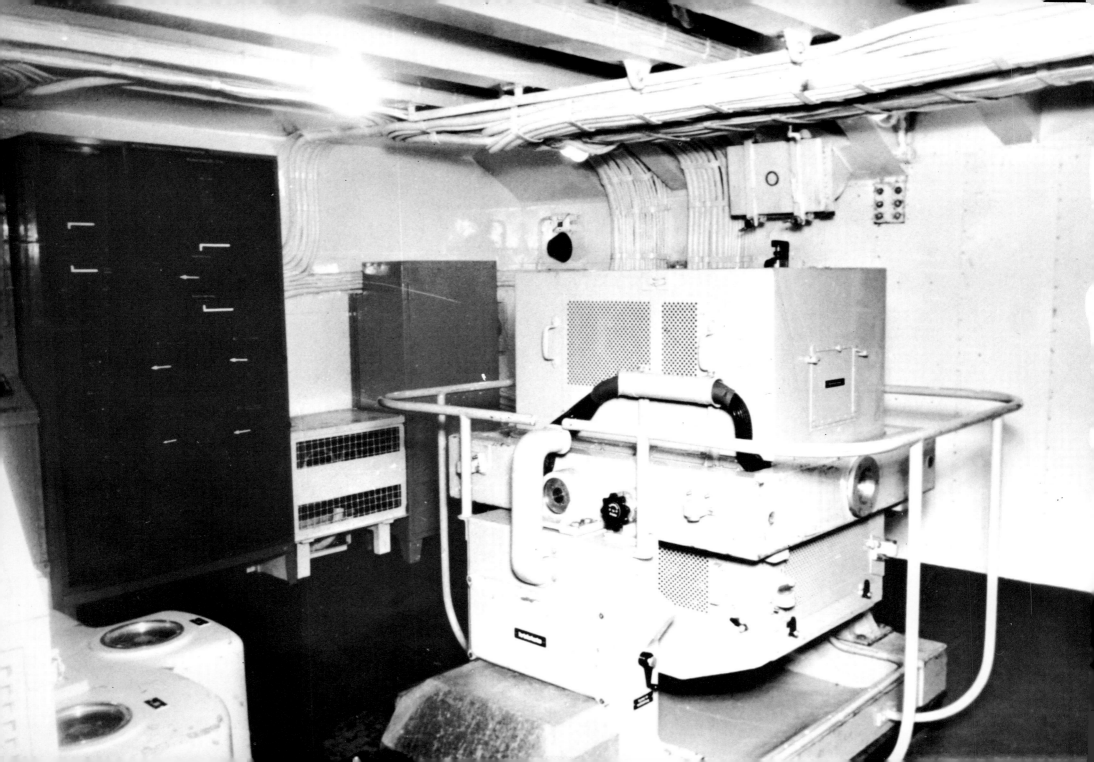

Left page and above:
In the stabilizer room with stabilizer and controls. By
trimming the water ballast and heating oil, the rolling
motions of the BISMARCK could be stabilized. The "sensor"
that controls the corrective measures is the stabilizer.

Among the deck equipment are many water hoses for cleaning
the deck and putting out fires; they are rolled on large drums.
At the lower left is the corner of the water connection.

Loudspeakers for normal on-board use are all over the ship (at
all stations). This loudspeaker can be protected from sea-water
damage by heavy flaps. Other loudspeakers can be seen in
other pictures.

Production 1939-43: 435 machines (most often-built German seaplane)—type of use: light fighter/on-board reconnaissance plane: the pilots were members of the Luftwaffe—Motor: BMW 132-K'812 HP—Dry weight: 2990 kp—Flying weight: 3730 kp—Top speed: 320 kph at 4000-meter elevation—Climbing speed: 300 meters—Range: 1070 operational ceiling: 7020 meters—Armament: 2 20-mm MG-FF (in the wings)—1 7.9-mm twin MG-81-Z—2 50-kg bombs—Crew: 2 men.

The catapult system shown here is a rotating catapult. A double catapult was installed on the BISMARCK. But both systems work by the same principle. The visible compressed-air container produces the energy needed for the process of acceleration, which is supported by the airplane's own power. At first the machine is braked off. At the highest resulting pressure and at maximum motor load, the catapult mechanism is then released, giving the machine the necessary takeoff speed.

After operations, the airplanes must land on the water near the ship and be lifted on board.

Six multipurpose airplanes of the type shown here were carried on board the BISMARCK.

The Ar(ado) 196—a heavily armed, robust low-wing monoplane with two pontoons (with wings folding onto the fuselage)—ranked among the most reliable German seaplanes.

The big airplane hangar, in which four planes are housed. For the sake of better space utilization, the wings of the Ar 196 are folded before the planes are pushed into the hangar. The heavy door, which is lifted sideward to the left, closes watertight. On the lower part of the door is an extra bulkhead. At the height of the seaman is the catapult, now covered by plates.

During the "Rheinübung" operation, nine airplanes, counting the three on the PRINZ EUGEN, will be available for fleet use. But none of these machines, though, will see action. Before the BISMARCK's last fight, its logbook is to be flown off. But because of damage to the catapult mechanism in the battle, this plan will have to be given up.

The catapult frame seen from above: This double catapult can be operated to both sides of the ship. To lengthen the catapult distance, the frame can be extended telescopically (telescopic catapult).

The picture at left shows almost the entire "airport facility": At right in the picture is the big airplane hangar with its door half open; the catapult frame is between the hangar and the funnel' under the floodlight platform is a work crane (there is also one on the starboard side) plus the two airplane-boat cranes. To the left of the funnel, but not visible in this picture, is a smaller hangar for one more plane, with another on the other side of the ship.

The catapult is extended. In the foreground are a cutter and a dinghy, lashed down on stocks.

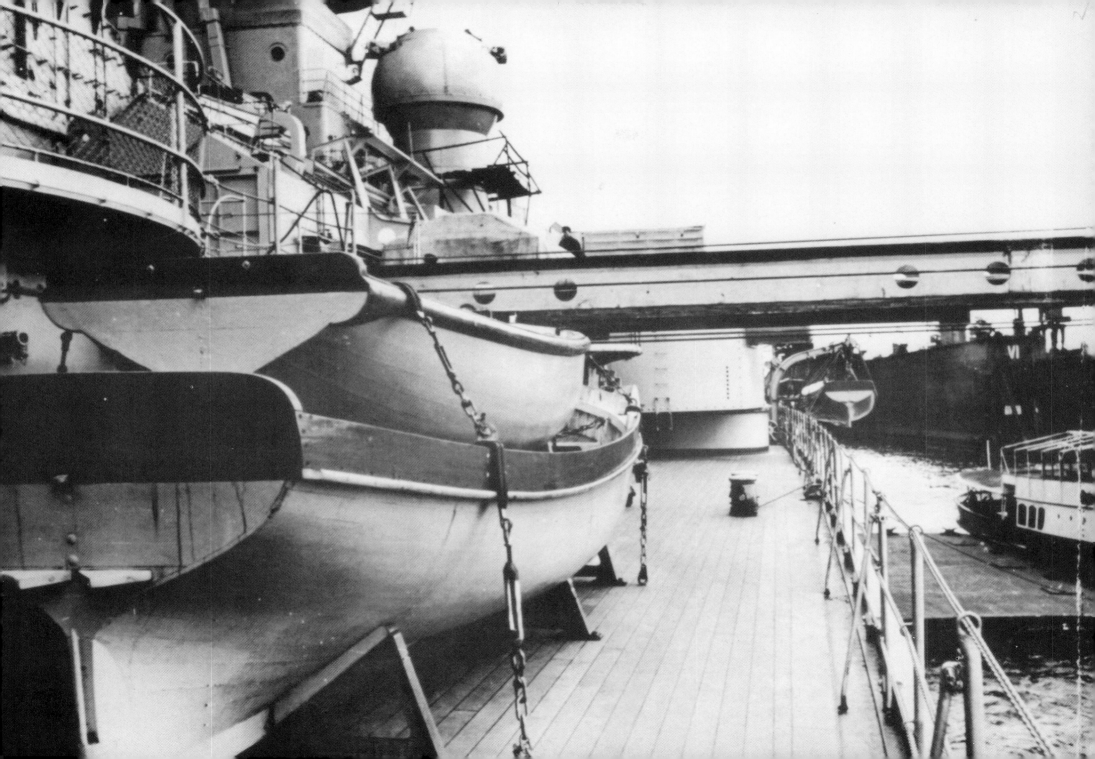

The pictures provide an impressive view of the technical complexity with which the lifting tackle was constructed. The more sensitive components are sea-watertight. The crane booms have a working length of about 20 meters (see also page 20).

A look aft on the rear boat deck. Behind the mainmast is one of the two anti-aircraft command posts installed on the afterdeck can be seen atop the deckhouse. The two "balconies" at right and left on the deckhouse are flag arms.

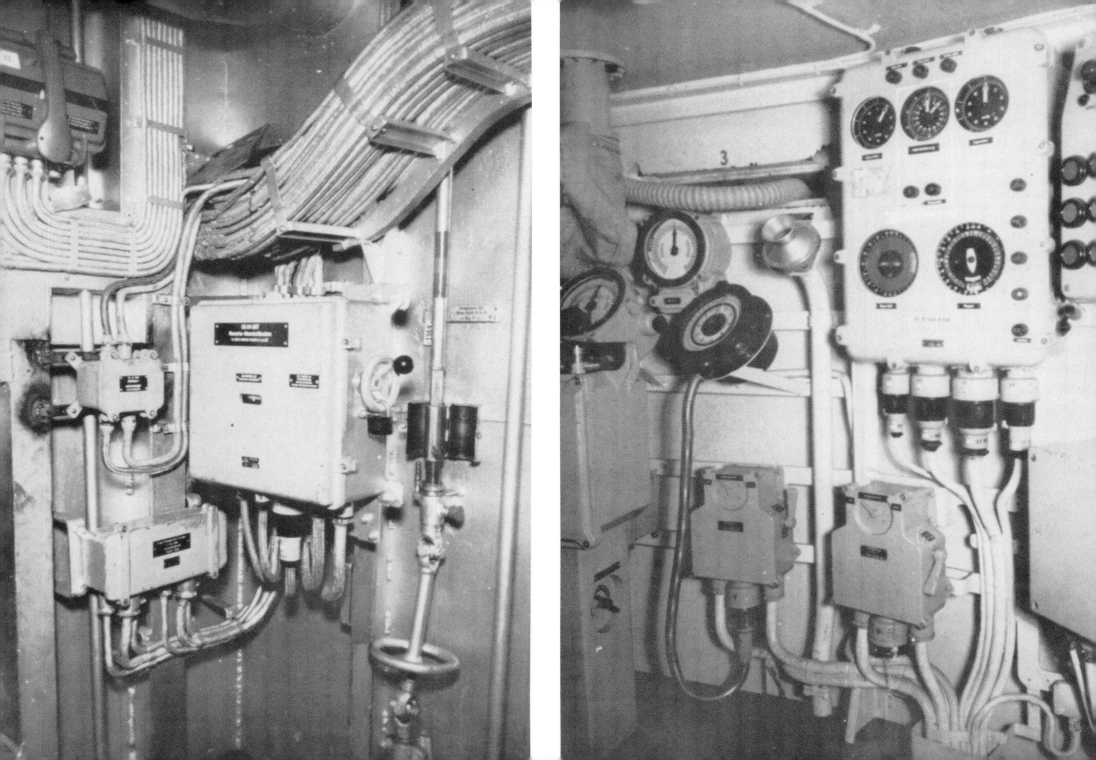

On the left page are two pictures of the machine control room.

Far left, the sea-damage switch box. In case of disturbances caused by seas or enemy fire, different power circuits can be switched on. The handwheel is secured by a lock. Only persons with special permission were allowed to use the switch wheel.

Near left, one of the control positions with speaking tube, repeater compass, rudder indicator and other instruments. Particularly noteworthy is the push-button switch box, a component that was rarely installed at that time.

One of the flag boxes, here on the deckhouse at the starboard side. Signal flags, each one with a definite meaning, are attached to the signal line. For many generations of mariners this was the only method of transmitting commands. On the BISMARCK this method had only a traditional character.

A group of air shafts, which supplied the various compartments of the "Berta" turret with fresh air. Such air shafts were all over the ship. When the areas they supplied were closed during operations, they were given excess pressure. The bent pipe in the center belongs to a ventilating duct.

Despite technical aids, it is still necessary to control modern battleships like the BISMARCK from the bridge. When in action, the bridge personnel should, of course, be protected in the armored command post, but certain stations, such as that of the target giver and advisor had to remain occupied. In addition, for reasons of long-standing marine tradition, the bridge is not to be deserted even then. Thus the bridge of the BISMARCK is equipped with command and control devices.

Visible in the picture is a rudder indicator (right). The gauges in the three armatures are rudder indicators for straight, steered and port courses. It provides, among others, the rudder operator as a monitor for rudder movement. Steering the ship is replaced by an automatic push-button steering apparatus. That is the rectangular box in the middle of the picture. The handle was held in both hands and the buttons pushed with the thumbs. The fact that a giant ship like the BISMARCK could be steered by push buttons is very impressive. A great deal of technology is required to turn pushing buttons into rudder movements.

The apparatus in the fore- and background are target-giver columns. The optical devices (between the forks) have not yet been installed. The target giver sits on a saddle-like seat and focuses the device on the target. The calculator automatically calculates the position of the ship in relation to its target. This information is passed on to the main computer. Finally, speaking tubes complete the equipment of the bridge.

The "black box" on the mainmast is a rudder indicator. In this case it shows a course straight ahead.

Providing daily necessities for more than 2000 men aboard the BISMARCK puts high demands on the ship's organization and the technical facilities, and requires practical supplying of needed articles. In addition, this part of the ship's operation must be carried out under limited space conditions and war conditions.

These pictures show come of the service areas of the BISMARCK.

Above: a storeroom for food at the height of rib 1664. On the ceiling is a conveyor for transport and lifting equipment.

Left: the mangle in the laundry.

The galley.

This and the following pictures were taken on Saturday, March 15, 1941, in the Scheer Harbor at Kiel-Wik.

The canteen. On the far wall is an advertisement for Bavaria-St. Pauli beer.

Left page: a look at the shoe repair shop.

Part of the bakery: The fresh vegetables are stored here only temporarily. The baking equipment is only needed when the ship is on the high seas. Until then, fresh bread is supplied from ashore whenever possible.

Work goes on constantly in the workshops, as here in the tailor's shop. A major warship is a unit of society, comparable to an army base. This work is done by the ship's personnel only in secondary roles. Primarily, they are fighting men with very definite military assignments when in action.

The ventilation system can be seen clearly in all the rooms.

The BISMARCK was planned and built for Atlantic operations. For these long war operations, not only hygienic and medical facilities but also "social rooms" had to be created.

The pictures show crewmen "chowing down" in their living quarters.

The daily life of the crewmen below decks was lived by artificial light. Living quarters with natural light were available only for the highest-ranking officers on board.

The picture above was taken when the BISMARCK was put into service on August 24, 1940. At the far left is a Blohm & Voss shipyard worker.

In the picture on the right page (also taken August 24, 1940): Mates in their non-commissioned officers' mess.

The rooms show varying furnishings. While the mates sit on upholstered benches, the seamen must make do with plain benches. The social level involved in the furnishings of the officers' quarters was even more striking. All the same, quite pleasant living conditions prevailed on the BISMARCK, compared to those found in small battle groups.

But in terms of the food served, there are no differences in either quality or quantity.

Here they are eating a casserole with bockwurst. It is March 15, 1941, a Saturday, and thus the typical German day for a casserole. The impression that these pictures give could be peaceful if one did not know that the year was 1941 and that these men had only a little more than two months to live . . .

The British Battleship HOOD— the BISMARCK's Great Opponent

A report about the BISMARCK would have been less than complete if one neglected to include the battleship HOOD in the portrayal, for at this time it was the British capital ship most closely comparable with the BISMARCK. To be sure, the BISMARCK could defeat the HOOD and was statistically somewhat stronger, but basically there is no superiority to be found in a comparison of the two ships' technical details. Also a factor, but not calculable for the fighting power of a warship, is, among other things, the level of training of its crew and the concepts of sea strategy by which the ships were built and operated.

The impetus to build the HOOD goes back to 1915. Originally designed as a battle cruiser, it was regarded, after various construction changes, as a fast battleship when it was put into service in 1920.

Even though its heavy artillery no longer completely met the demands of 1940, this ship still ranked among the more modern battleships at that time on account of certain revisions.

Until its sinking, the HOOD was the most powerful unit of the Royal Navy.

Displacement and Measurements (as of 1940)

Construction displacement	42,462 tons
Maximum displacement	48,360 tons
Length at construction waterline	258.4 meters
Overall length	262.2 meters
Greatest width	31.7 meters
Draught	7.8 meters

Armament

8 38.1-cm L/42 ship cannon in double turrets
12 10.2-cm anti-aircraft guns in double mounts
5 units of 12 anti-aircraft rocket launchers
2 each 53.2-cm over- and underwater torpedo tubes

Machinery and Performance (on test runs)

Power	(four propellers)
Speed	31.7 knots
Fuel capacity	4000 tons (oil-fired)
Range	4000 nautical miles at 10 knots

Armor

The armor plate of the HOOD was arranged according to modern points of view. The effect of shots that had their effect behind the outer skin was to be nullified in armored areas, without essentially influencing the ship's stability and firepower. Armored areas were built in along and at right angles to the ship's sides (segmented armor), using highly elastic armoring material.

Outward from about the middle of the belt armor down to the keel, the HOOD was additionally protected by a torpedo bulge.

Armor thicknesses (maximum)

Fronts of heavy gun turrets	381 mm
Barbette armor above the deck	305 mm
below the deck	229 mm
Upper deck	32 mm
Gun deck	25 mm
Armor deck	51 mm
Citadel armor	127 mm
Belt armor	305 mm

Aside from certain excesses of the times, the HOOD could serve as a model for most battleships built since the end of World War I. Her technical features were accepted and used until the end of World War II.

The picture shows the HOOD after a basic overhaul in 1929-30, with the classic features:

The placement of the heavy artillery near the center line in two double turrets, one set above the other.

High fighting top in tripod form; forward of it, the armored command post with fire control system, and one of the two funnels close to the fighting top.

The mainmast is shortened and serves only as an antenna carrier and lookout post.

The medium artillery turrets in side positions (the number of barrels was reduced to six between 1940 and 1941 in favor of anti-aircraft guns).

British battleship construction ended with the VANGUARD, which was scrapped in 1960 after only fourteen years' service.

The Heavy Cruiser PRINZ EUGEN— the BISMARCK's Little Brother

The heavy cruiser PRINZ EUGEN, whose name was so closely linked with that of the BISMARCK, was the heaviest and last German cruiser to be built. Ship experts call this ship the most beautiful warship ever built.

The PRINZ EUGEN was at that time the last link in a chain of development, in which the constructive possibilities were fully utilized. Whether an optimal warship was created thereby is partially debatable.

Launched August 22, 1938 at the Germania shipyards in Kiel; 14,240 tons; length 210 meters; width 21.8 meters; draught 5.8 meters. Top speed 34 knots with a power output of 132,000 HP from three turbines. Range 6800 nautical miles. Armament: Eight 20.3-cm guns, twelve 10.5-cm anti-aircraft guns, twelve 3.7-cm anti-aircraft guns in double mounts, up to 28 2-cm anti-aircraft guns, partially in quadruple mounts, and twelve 53.3-cm torpedo tubes in units of three. Three aircraft. Crew of more than 1500 men.

The fighting value of this type of ship, disregarding the armoring and a sufficient range, was intended for oceanic warfare by virtue of a high speed.

Therefore external dimensions, engine performance and crew strength were required which partially corresponded to the same values of battleships of the 35,000-ton class.

All in all, the "PRINZ" was a happy ship and was the only heavy unit of the German Navy to survive World War II with full fighting capability.

The "little brother" of the BISMARCK. This expression became a popular saying with little justification, unless the similar architecture had given rise to this formulation. The length or number of joint operations cannot be meant by it. It was even true that the "PRINZ", as ordered by higher authorities, was not allowed to stay with the BISMARCK during its last battle.

It is true that the "PRINZ", like the BISMARCK, belonged to the new generation of ships, and that the two ships showed some of the same features. First of all, the ships' architecture is very similar:

—Double turrets in medium and raised positions,
—One funnel, placed close to the fighting top,
—Airplane catapult between funnel and deckhouse,
—Location of the artillery positions, etc.

In addition, the "PRINZ" had the same elegant and graceful beltline as the BISMARCK and even included such interchangeable components as-

—Range finders for sea-target fighting,
—Spherical anti-aircraft control posts,
—10,5-cm double anti-aircraft mounts,
—Floodlights,

-to name just the particularly noticeable features above decks.

PRINZ EUGEN—View of the fighting top and funnel (each from port and starboard).

Recognizable are

—Radar antennas, here considerably more complex and standing out more than on the BISMARCK (see page 124).
This picture originates from a time when German radar development had progressed farther than in the days of the BISMARCK;

—2-cm anti-aircraft quadruple mount in the crow's nest near the funnel. Originally there was a floodlight with its protective cover mounted here. On the funnel surface (port side) the semicircular reinforcement to which the cover was attached is still visible.

—the spherical anti-aircraft control post is visible in the background;

—forward of it is the boat crane;

—in the foreground is the rear 10.5-cm double anti-aircraft mount (covered by a tarpaulin);

—behind the funnel are lashed-down life rafts;

—in the starboard pictures, the end of a pontoon of one of the airplanes and the catapult frame are just visible at the left edge of the picture. Unlike that of the BISMARCK, this is a rotating catapult;

—the covering hoods (in the foreground of the lower pictures) belong to the rear units of three torpedo tubes.

—finally, defensive buoys (lower pictures, inner corners).

The forward part of the ship with the two 20.3-cm double turrets for the ship's heavy artillery. Here too, the construction features are very similar to those of the BISMARCK. These similarities led to fateful mistakes by the British during the battles with the HOOD. At first the enemy battleships concentrated their firepower on the PRINZ EUGEN, and when the British recognized the confusion and finally changed their target to the BISMARCK, valuable time had already passed in which the BISMARCK could score its decisive hits on the HOOD undisturbed.

Building, Finishing and Testing

The BISMARCK at the equipping pier of the Blohm & Voss yards in Hamburg.

The battleship was at this stage of construction in May or June of 1940. The superstructure had been almost completely installed by this time.

The BISMARCK in about August of 1940, shortly before begin put into service. The artillery and anti-aircraft control posts have not yet been installed.

This and the following pictures give an impression of the commissioning ceremony, which was customary in the German Navy. The seamen stand in rows, and in the right foreground are Blohm & Voss shipyard workers who took part in the ceremonies.

Saturday, August 24, 1940, the day of commissioning. The commander of the BISMARCK, Captain Ernst Lindemann, is awaited.

The crew is in division formation: "Dress to the deck seam!"

The commander has just come aboard and takes the report of the officer of the watch—Left: The officers approach.

Captain Ernst Lindemann inspects the honor guard at the level of Turret C (starboard side); behind him is the First Officer, (I.O.) Frigate Captain Hans Oels, and next to him, with adjutant's braid, Lieutenant Commander Burkhard Baron von Müllenheim-Rechberg, the only officer of the BISMARCK to be rescued.

Then he inspects the officers. Of the 17 saluting officers, several can be identified (from left to right): Chief Engineer (LI) Corvette Captain (Eng.) Engineer Walter Lehmann; Ship's Administrative Officer Corvette Captain (V) Hartkopf; First Artillery Officer (I.A.O.) Corvette Captain Adalbert Schneider; Second Artillery Officer Frigate Captain Helmut Albrecht; Watch Officer/Division Officer Lieutenant Commander Rudolf Troll; Electrical Engineer Corvette Captain (Eng.) Freytag; Lieutenant Commander (Eng.) Jahreis (Leak Engineer/Turbine System); Anti-Aircraft Artillery Officer (Flak-A.O.) Lieutenant Commander Karl Gellert.

During the Captain's address on the afterdeck. The officers stand below Turret D.

Two signal mates at the flagstaff just forward of the flag tackle. One of the mates carries the flag, still rolled up, under his right arm. Appointment to such military acts of honor always was a reward for those chosen.

A few moments later:

"Hoist the flag!"

The **BISMARCK** is in service.

In navies at this time the flag was still more than a sign of national identity and tradition. For sea war, the waving flag was the sign of readiness to fight, regardless of the military conditions on the ship flying the flag.

After the commissioning:

"Dismissed below decks!"

The German battle flag was introduced on November 7, 1935 (Hitler: As of today I give the reactivated Wehrmacht of universal service obligation the new national battle flag. The Swastika shall be your symbol of the unity and purity of the nation, a symbol of the strength of the National Socialistic philosophy of life, the foundation of freedom and strength of the Reich. The Iron Cross shall remind you of the unique tradition of the old Wehrmacht, the virtues that inspired it, and the example that it set for you. To the national colors of black-white-red you owe your loyal service in life and death. It shall be your pride to follow the colors . . .")—The battle flag flew exactly 114 months aboard German warships (to May 8-12, 1945)—In the Wehrmacht the Swastika was officially introduced only on February 12 or 17, 1934; meanwhile the training cruiser KARLSRUHE had already left Germany in October of 1933 (!) for a long foreign cruise, flying the Swastika.

In September the BISMARCK was already leaving the shipyard (see left page). Shortly thereafter she sailed through the Kaiser Wilhelm Canal.

After being commissioned, hard work began at once to make the BISMARCK a powerful weapon of naval warfare.

In the same month, test runs were begun in the Baltic Sea (see illustration). The artillery control posts have not yet been installed. For some time after the war began, the Baltic Sea was an ideal testing and training area for the navy. Undisturbed by enemy aircraft or naval units, the necessary drills and tasks could be carried out here.

In December of 1940 the BISMARCK again passed through the Kaiser Wilhelm Canal on its way back to Hamburg, in order to have testing damage and deficiencies set right at the shipyard.

On Saturday, March 9, 1941, the battleship finally passed through the canal for the last time on its way to Kiel.

In Kiel, as subsequently at Gotenhafen, the last preparations for Operation "Rheinübung" were made. The level of training and the material equipment are completed and corrected. This included the application of camouflage paint and provisioning (all pictures taken in Kiel, March 1941.

Signals to the BISMARCK, which is putting out to sea. The battleship is now fully equipped.

TIRPITZ

On Saturday, April 1, 1939, the TIRPITZ followed its sister ship, the BISMARCK, into its element. At 52,600 tons, the newest German battleship was more than 2000 tons larger than the BISMARCK. Because of its heavier armament with light anti-aircraft guns, up to 64 2-cm barrels, and two sets of four torpedo tubes, it had a larger crew of 2608 men. Its range was increased to 9000 nautical miles. Otherwise the same statistics as those of the BISMARCK apply. The TIRPITZ was lost around 8:45 A.M. on November 12, 1944 in the vicinity of Tromsö, Norway, during an attack by 36 British "Lancaster" bombers, sinking with 1204 crewmen after taking direct hits and near misses.

As of March 1941, at which time the BISMARCK was preparing for Operation "Rheinübung", the German Navy possessed four battleships, the others being the SCHARNHORST and the GNEISENAU.

50

Adolf Hitler and the BISMARCK

On May 5, 1941 Adolf Hitler arrived in Gotenhafen. Aboard the fleet tender HELA he came to the Gotenhafen roadstead, where he spent some five hours. On the occasion of his visit, various battle drills and equipment demonstrations were carried out.

Adolf Hitler also visited the TIRPITZ. This picture was taken when he had left the BISMARCK and was aboard the HELA on his way to the TIRPITZ.

In the foreground, from left to right: (Presumably) Martin Bormann (NS-Reich Leader/Leader of the Party Chancellery/Reich Minister), Field Marshal Wilhelm Keitel (Chief of the Supreme Command of the Wehrmacht, OKW), Hitler and, at the far right, Flotilla Chief Admiral Lütjens.

On this occasion Hitler had a conference with Admiral Lütjens, who was to take command the group for Operation "Rheinübung." In the process, Lütjens informed the Führer as to details of the operation and also expressed his views on the threat of British torpedo planes, which could endanger the success of naval operations.

The Ship's Armament

Battleships were highly developed "weapon platforms", for whose construction all the technical potentialities of their times were fully utilized. All the characteristics of this type of ship, including its technical systems, essentially served in its task of using its weapons effectively against the enemy's large warships.

As development progressed, more and more guns of greater calibers and proportionally increasing barrel lengths were installed in battleships.

By the comparability of heat-technology processes in gun barrels and in heat-powered machines, the "performance delivery" of the two producers of energy can be compared. Thus the eight barrels of the BISMARCK's heavy artillery (SA) together produced some sixty million horsepower in a full salvo during the moment of full development for fractions of a second. The requirement for this was that the time of ignition in the loading chambers (combustion chambers) of all the guns would take place at the same moment.

Even though the chance of this happening was almost exclusively possible only in theory, it still had to be taken into consideration in the construction of large warships. It could be foreseen that the mechanical control of these forces and material characteristics would set limits to the building of guns.

One yardstick for it was, among others, the endurance of gun barrels, which in the BISMARCK added up to some eight "working seconds"; that is the sum total of the time in which shells pass through the barrel before it has to be replaced. While in England the arrangement of heavy artillery in triple (RODNEY) or even quadruple (PRINCE OF WALES) turrets was introduced for new battleships, the classic arrangement of the heavy artillery in double mounts was retained in the BISMARCK (and TIRPITZ). The technical advantages that one expected from triple mounting, such as concerted firepower and central ammunition supply, could not prevail against the disadvantages in the rough conditions of war, because, for example:

—the wider turrets required larger apertures in the deck. This detracted from the ship's static stability.

—the succession of firing was irregular and slower, because the central barrel could not be supplied with ammunition at the same time as the outside guns.

—the firepower could not be divided so well.

—if a turret was knocked out by enemy fire or mechanical problems, the total firepower was reduced considerably.

—the bigger, heavier turrets required larger sources of rotating power.

Caliber sizes, aside from a few exceptions (for example, the Japanese battleship YAMATO, with seven 46.0-cm guns), had settled on a diameter of about forty centimeters.

For the BISMARCK, the caliber of 38 cm for the heavy artillery was prescribed at the time of its planning. German industry had already been able to gather valuable experience with this caliber in the major warships of the Imperial Navy, and had already taken it into consideration in upgrading the artillery of the "Scharnhorst" class battle cruisers. Thus it was a natural result when, for logistical reasons, the BISMARCK was likewise armed with guns of this caliber.

Unlike British battleships of the same generation, the BISMARCK was particularly noted for being strongly armed with medium artillery (MA). Since this German warship had been built for Atlantic operations, and thus could not be given and light cruisers and/or destroyers for support, what with the strategic geographical situation of the German Navy, the BISMARCK really had to be armed with weapons of this caliber.

Since the German warships, for lack of suitable ships, had to do without the support of on-board aircraft, attempts were made to make up for this lack with effective anti-aircraft armament. The mounts of the heavy anti-aircraft guns were stabilized with yawing, rolling and pitching axles, meaning that the ship's movements were compensated for by technical devices.

Along with its fire control systems (see page 124), the BISMARCK was the best-equipped ship of its time in terms of artillery:

—the turrets and gun mounts represented the state of the art in terms of construction.

—this resulted in a fast firing speed,

—a saving of weight for the benefit of other facilities, and

—better firing accuracy.

—the explosive charge of the German shells amounted to about 2.5 times that of the British.

It would not be right, though, to make the assumption that the British shipbuilding industry had been less capable than the German. Because of the naval-political pressure on the British Navy, the requirements for quality had to be subordinated to what was necessary for the ships.

In the process, the British built so cost-effectively that for what the BISMARCK cost, the British could have built two battleships which, together, surely could have had the same firepower plus a simultaneous increase in the individual ships' potential.

In general, though, improvements in quality cost progressively more from a certain technical borderline on, but no longer result in notable increases in fighting value. In the BISMARCK some technical advantages were not used for reasons of construction, which meant that limiting the barrel elevation of their heavy artillery cost them their superior range in comparison with the new 35.6-cm shells of the "King George V" class battleships.

Previous page: This picture shows clearly how the heavy artillery turrets have determined the architecture of the BISMARCK. When one realizes that the 15-cm double turrets (visible roughly amidships), weighing over 60 tons in and of themselves, were among the main armament of light cruisers, the relative sizes make clear the effect of the heavy artillery.

Weapon type, number, caliber	Year built	Barrel length in caliber lengths	Barrel weight (tons)	Shell weight (kp)	Muzzle velocity (meters per second)	Firing speed (shot per min. and barrel)	Ammunition supply per gun	Shot range at max. barrel elevation (hm/degree)	Barrel endurance in number of shots	Location
Heavy artillery (SA) 8 x 38.0	1934	L/47	1092	798	850	3	105-120	362/35.0	200	Always in two double turrets, forward and aft, two of them raised
Medium artillery (MA) 12 x 15.0	1928	L/55	9.1	43.3	850	10	150	220/35.0	2500	Always in three double turrets, port and starboard
Anti-aircraft guns										
16 x 10.5	1932	L/65	--	15.1	880	12-15	420	177/44.0 elevation 12,800 m	3500	Always in four double mounts on the boat deck, to port and starboard
16 x 3.7	1930	L/83	0.12 weight of double mount 3.670	0.745	1000	ca. 30 semi-automatic	2000	85/44.0 elevation 6800 m	--	8 double mounts on the upper deck, bridge structure and crow's nest of foremast
12 x 2.0	1930	--	--	--	840	theoretically 200 fully automatic	--	effective shot height ca. 400 m	--	mounted singly

Turrets "Anton" and "Berta." August 1940—Blohm & Voss.

September 1940 in Kiel Bay: the BISMARCK "points" the barrels of the heavy anti-aircraft guns.

The arrangement of the armor plates around the turret area is very easy to see here. The greatest armor thicknesses are united here. The front wall, just visible here, has a thickness of 360 mm; the smooth side wall is 220 mm thick and the slanted side wall is 150 mm. The visible part of the barbette is 340 mm above the deck and 220 mm below.

The turret socket of a 38-cm turret hangs on the hook, ready for installation. The lower part is set on the turning ring. A gear drive in the socket housing turns the turret in the turning ring. The entire weight of the turret moves on ball bearings. The turning ring's diameter between the bearings measures 8780 mm. The picture also shows the electric motor for the turning drive at left. The traverses that are visible on the socket platform carry the gun cradles. The front part of the turret is part of the picture on the right page. About one-third of the socket rises above the deck and is protected by a ring armor (barbette).

Obviously, this part is the bedding ring for one of the 38-cm turrets. This ring was installed on the deepest part of the turret shaft. This part of the turret system is called the shot platform. The shells are moved out from here. Over the shot platform is the cartridge platform. Because of the great weight, shells and cartridges must be loaded separately.

View over the forward fighting command post, the bridge and the forward heavy artillery turrets. The visible cover of the command post has an armor thickness of 200 mm. The numerous periscopes for sea- and air-space observation, for use in battle, are striking.

Shift change at Blohm & Voss. The "Cäsar" turret aft is visible, the barrels at different elevations. The guns were loaded at 2.5-degree elevation (loading position); they could be adjusted between -5.5 and +35 degrees. For reasons of construction, the full shot range at 44 degrees elevation was eliminated. Otherwise the base, about five meters long, measured from the shield gudgeon, would have extended even farther down into the turret shaft, and the machine platform would have had to be set deeper (starboard side).

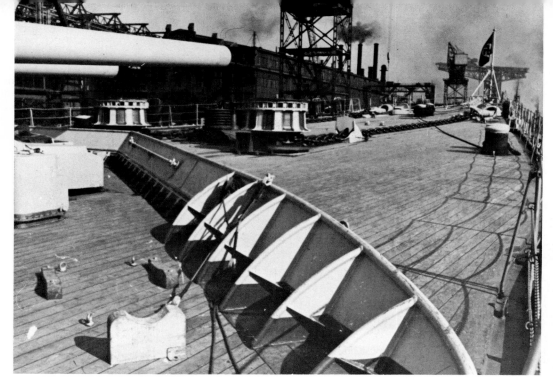

View toward the bow, showing the forward gun barrels. The ribbed raised panels that can be seen all over the ship are wave breakers intended to hold back water that comes onto the deck. The barrels have a length of 47 caliber lengths. If one multiplies the caliber size by the caliber lengths, one gets the barrel length, about 18 meters on the BISMARCK.

The mouth of a 38-cm barrel.
The 38-cm caliber is based on the British caliber of 15 inches and had to be adapted to the decimal system at about 38.1 cm by this standard. But since the Germans use the decimal system, it is correct to classify this caliber as 38 cm.

Left page:
The barbette armor of the "Berta" turret rises as high as a house above the maindeck. Air intake shafts have been attached to the cylindrical wall. Between the two heavy artillery turrets, part of a 15-cm double pivot mount rests before being installed. At right of the locking block, the elevating ring cab be seen.

The barrels of the heavy artillery reach out over the edge of the hull. It almost seems as if this turret position would influence the sea stability of the ship. The muzzle brakes and runners of the mountings were so excellently balanced against the weight of the barrel recoils that even a full salvo did not noticeably influence the ship's movements. The attachment of heat-protection covers is noteworthy. These were meant to avoid one-sided cooling or heating of the barrels, in order to minimize the ballistic influences that would have affected the accuracy of the shot.
On the pier are two mates (gun leaders)—Kiel, March 1941 (Scheer Harbor).

Here a 15-cm gun turret is just being set on the clearly visible turning ring. The ball bearings are equally easy to see. If one uses the size of the hand at the lower edge of the picture as a basis of comparison, one gets an approximate impression of the size of the bearings. The diameter between the bearings is 3630 mm. The barbette armor has a thickness of 50 mm. From the rim of the barbette, the turret shaft goes down almost eight meters into the ship's hull.

The picture on the right page shows one of the two medium double turrets. The muzzles point in the direction the ship is going. The box extension on the left side of the turret is also present on the right side and holds a range finder, with which only the medium turrets were equipped. In addition, the picture shows two double mounts of the heavy 10.5-cm anti-aircraft guns.

A fully installed 15-cm turret. The elevation range of the guns was between -10 and +35 degrees. The cylinder sponson on the turret roof is the periscope protector. When in action, this protector is removed. When the BISMARCK was put into service, this turret system was no longer up to date. The Royal Navy and, in particular, the U. S. Navy replaced medium artillery with multipurpose artillery which could also be used against aircraft. The favorable calibers proved to be diameters around 12 cm, which have been retained, along with other calibers, to this day.

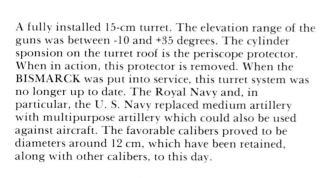

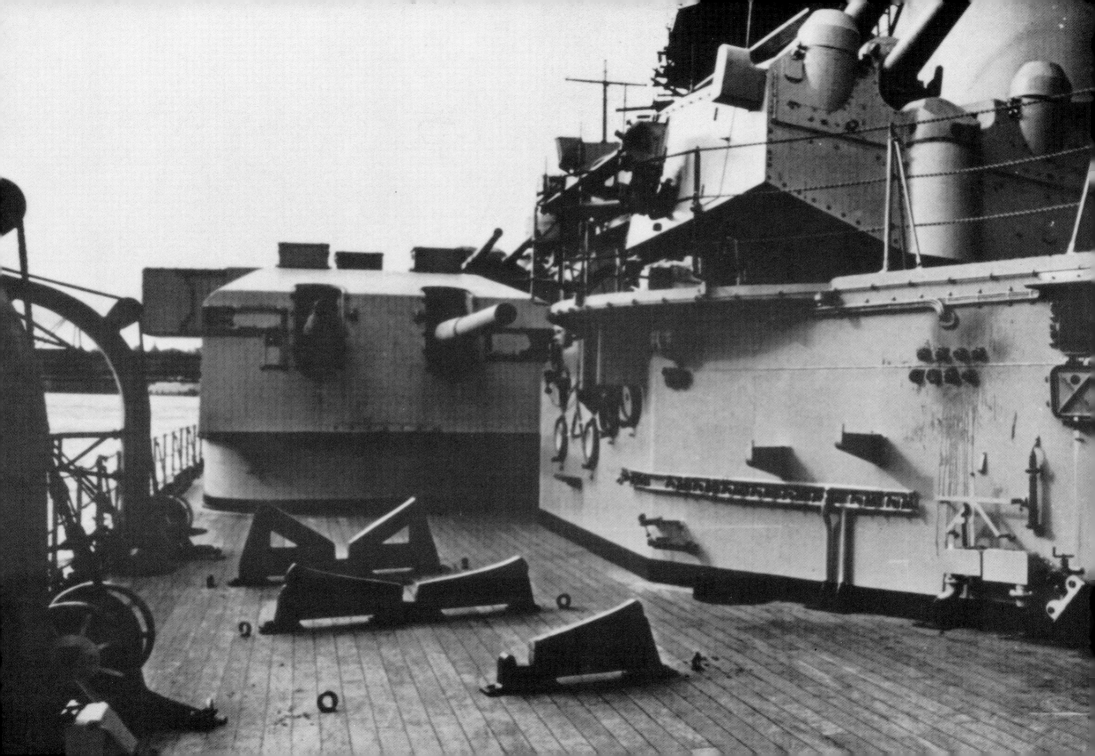

Left: This device is certainly a means of defense. At least six of these devices were carried on the upper deck. The significance and purpose of these devices could not be determined with certainty. They may have been anti-mine devices or sound buoys to defend against acoustically steered torpedoes. In any case, one must realize that the BISMARCK was not suited for minesweeping because of its deep draught. Acoustic underwater weapons were still unknown when the BISMARCK was put into service.

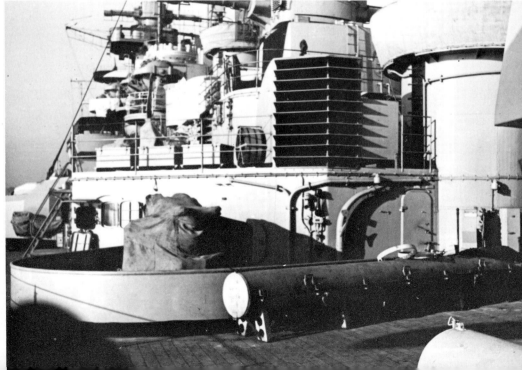

One of the heavy 10.5-cm anti-aircraft guns in a double mount at maximum elevation. At this elevation the guns could not be loaded. The firing speed could be kept very fast because, except for the loading process, all the other movements were fully automatic. As can be seen in the picture—the two naval anti-aircraft gunners wear headsets—the fire-control settings and commands were transmitted from a fire-control headquarters (one forward, one aft) to the gun crews (starboard side).

Left page: View forward from aft (port side) at the level of the "Cäsar" turret: in the foreground is a container for the barrel-cleaning tools and, covered by a tarpaulin, twin 3.7-cm anti-aircraft guns.

This picture shows clearly how the 10.5-cm anti-aircraft guns are stabilized. On the front wall of the mantlet is the pivot column for the rotating movement (yawing axle). The actual "mantlet box" is mounted so that it can turn perpendicularly to the pivot column, and could "roll" (rolling axle) in the distance between the outer edge of the mantlet and the ship's deck (see the picture). The pitching action of the ship is equalized by the elevating ring.

With this equipment, the ship can move in any direction under the mantlet without the gun barrel's position changing in regard to the target.

The optical glasses at left and right near the gun barrels belong to the optical aiming system used in sea-target fighting.

These weapons systems are so heavy that they can only be installed on major ships. At the war's end their performance was no longer sufficient to attack warplanes effectively at the highest attainable altitudes.

Between the house-high parts of the ship, the light anti-aircraft guns are not especially visible. These pictures show very clearly how they are set up.

Upper right: The forwardmost 3.7-cm twin anti-aircraft gun on the upper deck between turret "Berta" and the bridge. Underneath, on the main deck, the single mantlet of a 2-cm gun can be seen. The gun crews are just being introduced to the equipment, although the shipyard personnel are still working on board. To shorten the training time on board, the technical and weapons personnel were acquainted with the technical equipment of a ship even before it was put into service, when possible (port side).

Below: The after 3.7-cm anti-aircraft position. The crew is not protected by shields.

Right: On the floodlight platform behind the tarpaulin, the barrel of a 20cm anti-aircraft gun can just be seen. At left on the platform is the ammunition box for ready ammunition (port side).

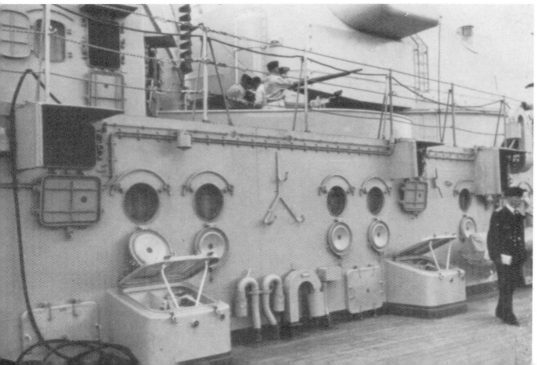

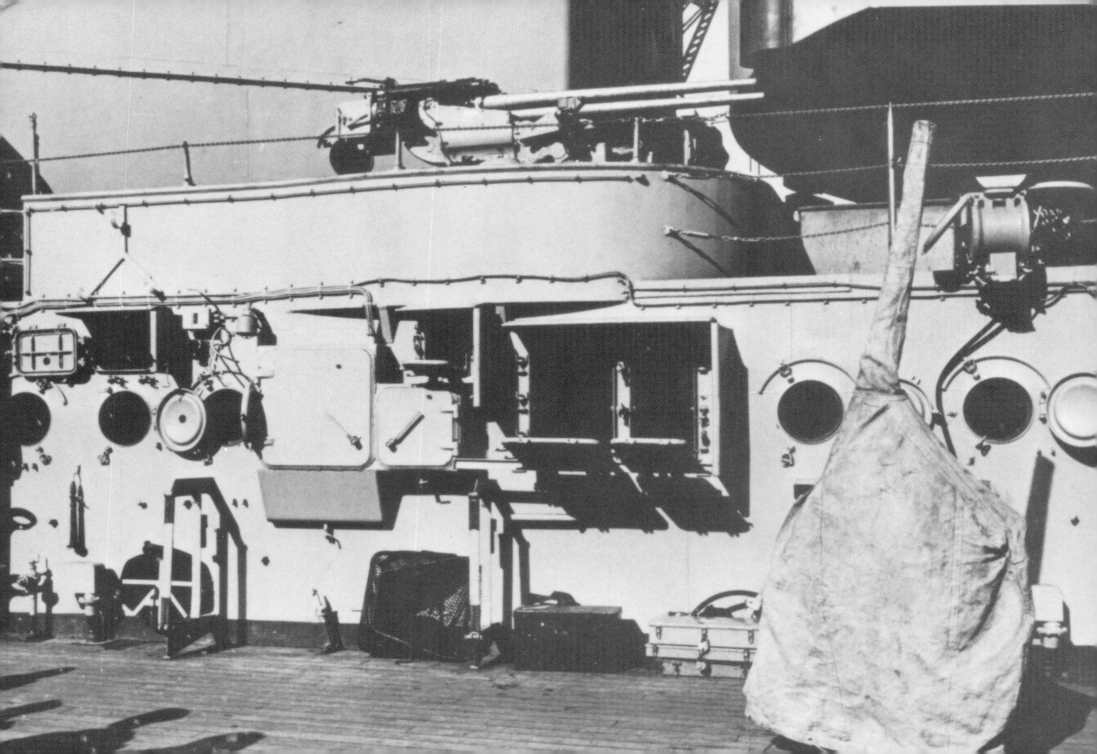

The forward 3.7-cm port anti-aircraft position and a covered 2-cm anti-aircraft gun. The turning ring-socket mantlets of the 3.7-cm guns are likewise stabilized by three axles, and a barrel length of 86 caliber lengths makes it a precision weapon with which aerial targets can still be attacked at medium altitudes. The arc of elevation lay between -10 and +75 degrees. The system had the disadvantage of being only semi-automatic and thus not having a very fast rate of fire. But in navies it was still thought at that time that this technical weakness could be compensated for by precise weapon and mantlet construction and great shot elevation, with a resulting saving in ammunition. Events of the war very quickly changed this view, which resulted in the obsolete 3.7-cm anti-aircraft guns being replaced by weapons with better performance, such as the 40-mm "automatics" by Bofors.

The 2-cm anti-aircraft machine gun is, of course, a very modern weapon, but the twelve of them aboard the BISMARCK were simply too few. Only when they were installed in sets of four on other ships from 1941-42 on did the fire become heavy enough. All in all, the equipping of the BISMARCK with light anti-aircraft artillery is no longer up to date.

The heavy anti-aircraft guns are unusually heavily concentrated amidships.

Kiel-Wik (Scheer Harbor), March 1941.

Next page: The British called the BISMARCK "a symbol of the Third Reich's sea power."

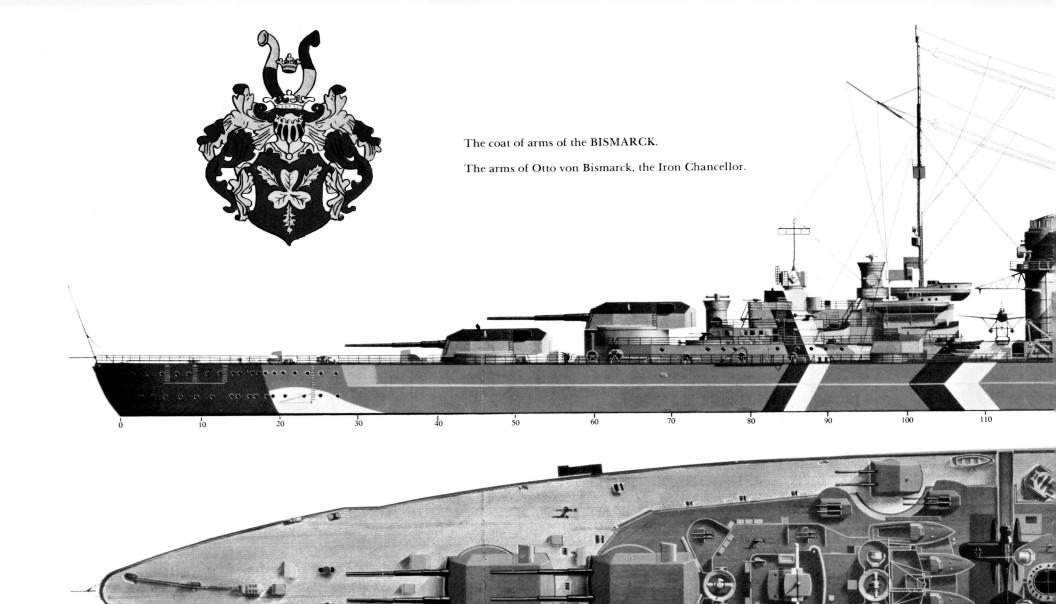

The coat of arms of the BISMARCK.

The arms of Otto von Bismarck, the Iron Chancellor.

70

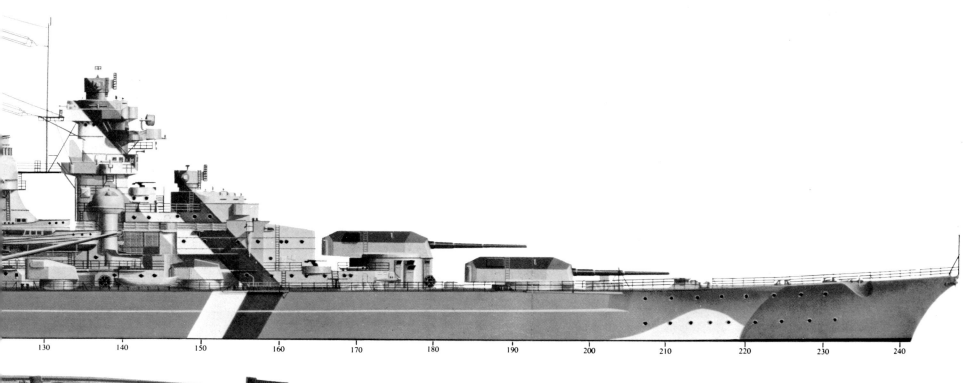

130   140   150   160   170   180   190   200   210   220   230   240

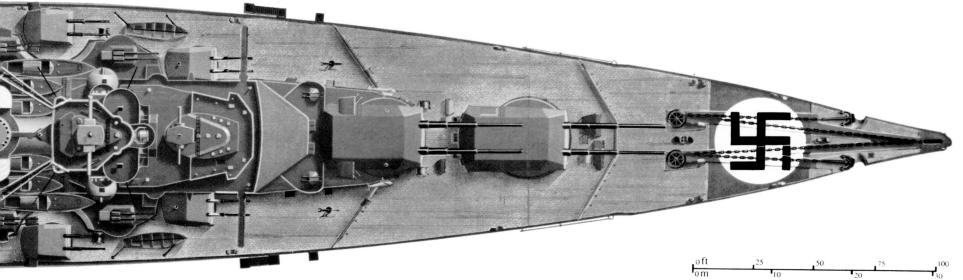

0 ft   25   50   75   100
0 m   10   20   30

From Gotenhafen to the Denmark Straits

Operation "Rheinübung"

The Operation "Rheinübung" was planned with the goal of disturbing or even crippling shipping on the main Atlantic routes for the duration of the operation. Combat with enemy over-sea forces was thus to be avoided if possible. Today such operations would be referred to as "hit and run."

The German naval command hoped to be able to follow up the success of the SCHARNHORST-GNEISENAU operations, which together had sunk 22 ships totaling 115,600 tons between January 22 and March 22, 1941. The tactical gains could be evaluated just as highly as the material success, in that such heavy naval units could obviously operate in the Atlantic almost without disturbance.

On May 18, 1941 in Gotenhafen, a few hours before the beginning of Operation "Rheinübung", the Fleet Chief, Admiral Lütjens, visited the heavy cruiser PRINZ EUGEN and its crew. This constituted an act of protocol and was also one of the few opportunities for an officer at Lütjens' level to come into contact with the lower service ranks at all. The separation of commanding officers from their underlings was especially marked in the navy and results from the particular nature of naval warfare.

Admiral Lütjens inspects the line of mates (non-commissioned officers); at the left of the Fleet Chief is the officer of the watch, at right is PRINZ EUGEN Commander Brinkmann with the crew roster in his hand, as well as the First Officer (I.O.), Frigate Captain Otto Stoos.

Unlike the BISMARCK, the PRINZ EUGEN at this time already had put in almost two years' service. The crew was excellently trained and accustomed to each other.

After the inspection, just at 11:12 A.M. on the same day, the operation began. With the help of tugboats, the BISMARCK left the harbor (right).

In Command of the BISMARCK

(crosses after names † casualties; only one officer survived)

Commander:
Captain Ernst Lindemann†
Knight's Cross of the Iron Cross: 12/27/1941

First Officer (I.O.):
Frigate Captain/Captain Hans Oels†

Navigation Officer (N.O.):
Corvette Captain Wolf Neuendorff†

First Artillery Officer (I.A.O.):
Corvette Captain Adalbert Schneider†
Knight's Cross of the Iron Cross: 5/27/1941 (!)

Second Artillery Officer (II.A.O):
Frigate Captain Helmut Albrecht†

IV. Artillery Officer (IV.A.O.)—Adjutant to the Commander:
Lieutenant Commander/Corvette Captain Burkhard Baron von Müllenheim-Rechberg (only surviving officer)

Anti-Aircraft Artillery Officer (Flak-A.O.):
Lieutenant Commander/Corvette Captain Karl Gellert†

First Watch Officer (I.W.O.)—Division Officer:
Lieutenant Commander Rudolf Troll†

Chief Engineer (L.I.):
Corvette Captain (Eng.) Engineer Walter Lehmann†

The Fleet Staff:
As of May 1941 the BISMARCK succeeded the GNEISENAU as the new flagship of the fleet.

Fleet Chief:
Admiral Günther Lütjens†
Knight's Cross: June 14, 1940.

Chief of Staff:
Captain Harald Netzband†

First Admiral Staff Officer (1. Asto/AI):
Captain Paul Ascher†

Second Admiral Staff Officer (2. Asto/A II):
Captain Emil Melms†

Fourth Admiral Staff Officer (4. Asto/A IV):
Corvette Captain Hans Nitzschke†

Other Officers of the BISMARCK:

First Lieutenant Jürgen Brandes†
Naval Senior Staff Surgeon Dr. Busch†
Lieutenant Doelker†
Dr. Externbrink (Meteorologist)†
Corvette Captain (Eng.) Freytag † (Electrical Engineer)
First Lieutenant (Eng.) Giese†
Lieutenant Rolf Hambruch†
Corvette Captain (V) Hartkopf†
(Ship's Administrative Officer) (VO)
First Lieutenant Friedrich Heuser†
Lieutenant Commander (Eng.) Jareis (Turbine System)
Lieutenant Commander (Eng.) Engineer Gerhard Junack†
(Central Turbine Room)
First Lieutenant Kardinal†
Lieutenant Commander Krueger †
(Artillery Calculating System)
First Lieutenant Hans-Gerd Lippold†
Lieutenant Commander Mihatsch† (or Michatsch)
First Lieutenant (Eng.) Richter†
Corvette Captain Max Rollmann†
Naval Architect Schlüter†
Lieutenant Schmidt†
First Lieutenant Sonntag† (VI. Division)

Operation "Rheinübung"—Participating Units

The operation required not only careful planning but also high material and logistical costs. Whereas British ships could operate within range of supply bases, German warships on long-range operations had to be supplied and maintained at sea. Nineteen supply and other service ships were available for this purpose, spread around the planned area of operations according to a well-thought system. After the sinking of the BISMARCK, the whole support network was disposed of by the British, more or less as a result of treason.

For reconnaissance in the operations area, sixteen U-boats had been stationed there. Several boats regularly carried out search missions together. But they were positioned so badly that they could not hold up the approach of the enemy ships against the BISMARCK. It cannot be ruled out that the British Navy had broken the German Navy's code, so that the British ships could be directed through the area threatened by U-boats.

The three destroyers shown here escorted the BISMARCK group until the latitude of Trondheim, where they were released on May 22.

All participating units are illustrated in the following pages.

Fleet Flagship: Battleship BISMARCK	Fleet Chief Admiral Günther Lütjens
	Commander Captain Ernst Lindemann
Heavy Cruiser PRINZ EUGEN	Captain Helmuth Brinkmann
Destroyers: Z-10 (HANS LODY)	With the Chief of the 6th Flotilla:
	Frigate Captain Alfred Schulze-Hinrichs
	Commander: Frigate Captain Werner Pfeiffer
Z-16 (FRIEDRICH ECKOLDT)	Frigate Captain Alfred Schemmel
Z-23	Frigate Captain Friedrich Böhme
Boats of the 5th Minesweeping Flotilla	Flotilla Chief: Corvette Captain Rudolf Lell
Outpost boats (10th VP Flotilla)	VP-1011, VP-10..
	VP-10..
Barrage breakers	13 and 31
Submarines	

U 46 (Endras)	U 98 (W. Schulte)
U 48 (H. Schultze)	U 108 (K. Scholtz)
U 66 (Zapp)	U 138 (Gramitzky)
U 73 (Rosenbaum)	U 552 (E. Topp)
U 74 (Kentrat)	U 556 (Wohlfahrt)
U 93 (Korth)	U 557 (Paulshen)
U 94 (Kuppisch)	BARBARIGO
	Ghilieri

Other Units/Auxiliary Ships

ALSTERTOR—Supply ship	HEIDE—Supply tanker
AUGUST WRIEDT—	ILL—Supply tanker (planned)
Fishing/Weather ship	KOTA PINANG—Scout ship
BABITONGA—Blockade breaker	LAUENBURG—
BELCHEN—Tanker	Fishing/Weather ship
EGERLAND—Tanker	LOTHRINGEN—Tanker
ERMLAND—Tender	MUNCHEN—Fishing ship
(planned, not present)	SACHSENLAND—
ESSO HAMBURG—Tanker	Weather opservation ship (WBS-7)
FREESE—Fishing/Weather ship	SPICHERN—Tender
FRIEDRICH BREME—Tanker	WEISSENBURG—Tanker
GEDANIA—Tanker	WOLLIN—Tanker
GONZENHEIM—Scout ship	

Right page: Destroyer FRIEDRICH ECKOLDT (Z-16)—Blohm & Voss, Hamburg—Launched: March 21, 1937—Commissioned: August 2, 1938—Fate: Sunk December 31, 1942, 11:55, in the Barents Sea in artillery action against British cruisers (SHEFFIELD, JAMAICA)—2239/2619/3165 tons, 36.0-38.2 knots—Armed: 5 12.7 cm, 4 3.7 cm, 8 TR 53.3 (16); 60 mines—Crew: 10 + 315. Commander: Frigate Captain Schemmel; I.O.: Lieutenant Commander G. Bachmann—Date of photo: 1938.

Above: Destroyer Z-23—Deschimag, Bremen—Launched: December 15, 1939—Commissioned: September 15, 1940—Fate: Sunk August 21, 1944 by British aerial bombs near La Pallice—2603/3605 tons, 36.0/37.5 knots—Armed: 4 15.0 cm, 4 3.7 cm, 8 TR-53.3 (16), 60 mines—Crew: 11 + 321.

Commander: Frigate Captain F. Böhme; I.O.: Corvette Captain Baron Freytag von Loringhoven.

Below: Destroyer HANS LODY (Z-10)—Germania Shipyards, Kiel—Launched: May 14, 1936—Commissioned: September 17, 1938—1945: Captured by the British (scrapped 1949 in Sunderland)—2270/3190 tons, 36.0/38.2 knots—Armed: 5 12.7 cm, 4 3.7 cm, 8 TR-53.3 (16), 60 mines—crew: 10 + 315.
Commander: Corvette Captain Werner Pfeiffer; I.O.: Lieutenant Commander Barkow—Date of photo: 1938.

When this picture was taken, the operation was only a few hours old. Bright sunshine and the peaceful surroundings of the Baltic Sea put the crew in an optimistic mood. For a short time the battleship TIRPITZ was still with the group. Maneuvers were carried out together. The meeting of these three ships formed the strongest concentration of German warships during World War II.

The British Captain Henry Denham (photo), then naval attaché in Stockholm, obviously played an important role in Operation "Rheinübung." Through his good contacts, he learned that the BISMARCK-PRINZ EUGEN group had passed through the Great Belt in the night of May 20. He was recognized by the British Admiralty for this. His report is said to have caused the British Admiralty to react for the first time. Independently of this, the German group was also sighted and reported by numerous foreign merchant ships and fishing boats (see picture above). Among them was also the Swedish flight-deck cruiser GOTLAND, which reported its sighting at once.

Left page:
This picture was probably taken on May 20 too. In the center of the picture is the BISMARCK, in the left background is the heavy cruiser PRINZ EUGEN, and at the edges of the picture in the foreground are two minesweepers of the Fifth Naval Scout Flotilla. It is not known why the group, for reasons of secrecy, did not pass through the Kaiser-Wilhelm-Canal separately. Surely this would have avoided the Coastal Command's sending out "Spitfire" reconnaissance planes to look for the BISMARCK and the PRINZ EUGEN around 11:00 A.M. on the very next day, May 21, 1941.

On the very same day the group was protected additionally by air security.

On May 21, 1940, between 7:00 and 8:00 A.M., the group neared the rocky coast of Norway.

About 8:00 A.M. (upper left) the group entered the fjord. Between 8:30 and 9:00 (lower left) the ships moved slowly into the inner area of the fjord. Shortly thereafter the group separated (below and right page). About 11:00 A.M. the BISMARCK anchored in the Grimstadfjord and the PRINZ EUGEN with the destroyers in the Kalvenes Bay.

At their anchorages, both ships changed their camouflage paint. The striped camouflage paint was painted over with a single color. Because of its shorter range, the **PRINZ EUGEN** also had to refuel.

To what extent the "zebra" camouflage was really able to fool attackers is debatable. Often the visual effect of this type of camouflage was accentuated by certain light conditions. The characteristic lines of the ships could not be hidden from aerial reconnaissance. By 1:15 P.M., just a few hours after anchoring, the ships had already been discovered and photographed by the reconnaissance planes of the Coastal Command, despite heavy German air-space securing. And just a little over two hours after that, Admiral Tovey, Commander in Chief, had information about the composition and anchorage of the German group. So as early as the middle of 1941, German fleet movements within the range of reconnaissance planes could no longer be kept secret.

The light ships stand out clearly from the dark waters in this excellent air photograph, and the hull of the **BISMARCK** looks very different from those of the freighters.

On the very same day, the BISMARCK and the PRINZ EUGEN left Kalvenes Bay at cruising speed, escorted by the three destroyers, on a northward course. The picture was taken when the BISMARCK passed the heavy cruiser about 9:00 P.M. and took the lead. The different camouflage paint is noteworthy; only the white wave near the bow remains.

Thursday, May 22, 1941. The three destroyers were released. The group now held a cruising speed of 24 knots, led by the BISMARCK.

About noon the weather conditions worsened; it became hazy and rainy. The speed could still be maintained. During the afternoon the visibility had worsened more, amounting to only 3000 to 4000 meters now. In the thickening fog, the BISMARCK had turned on one of its rear floodlights as a fog light for the following PRINZ EUGEN. On the PRINZ EUGEN, the anchor chains had been lashed down for the sea voyage. One can see by the lack of a bow wave that the speed had been reduced sharply now (lower left picture). At this time the battle group was already just short of the Denmark Straits, the northernmost sea route to the North Atlantic.

A few hours later, pieces of ice floated past the ships' hulls (below and right page—drawing by Walter Zeeden). On Friday, May 23, about 6:30 P.M., the group had reached the ice limit. This northernmost route through the Denmark Straits had to be chosen to circumvent minefields farther south. Now it was barely an hour before the German ships were spotted by a British warship. At about 8:30 the BISMARCK fired the first shots in Operation "Rheinübung."

Chronological Details of Operation "Rheinübung", May 18-23, 1941.

Sunday, May 18, 1941:

Between 11:12 and 11:15 A.M.:	BISMARCK and PRINZ EUGEN leave their berths in Gotenhafen and anchor in the roadstead.
11:30 A.M.:	Operation "Rheinübung" officially begins.
Afternoon:	Anchor raised: maneuvers with TIRPITZ near the coast.
Evening:	BISMARCK and PRINZ EUGEN anchor in Gotenhafen roadstead.

Monday, May 19, 1941:

About 2:00 A.M.:	Anchor up: separate cruise westward—subsequent meeting at Cape Arkona (Rügen) and cruise on to the Great Belt.

Tuesday, May 20, 1941:

About 2:00 A.M.:	BISMARCK, PRINZ EUGEN, and the three destroyers Z-10, Z-16 and Z-23 pass through the Great Belt.
1:00 P.M.:	German air security . . . the Swedish flight-deck cruiser GOTLAND (Agren) is sighted to the east—GOTLAND reports sighting BISMARCK . . .
Between 2:00 and 6:00 P.M.	:BISMARCK and PRINZ EUGEN are sighted or observed by numerous merchant ships and fishing boats.
May 20-June 1, 1941:	Operation "Merkur": German airborne landing on the island of Crete.

Wednesday, May 21, 1941:

6:40 A.M.:	PRINZ EUGEN B-service decodes British FT: Aircraft are to keep watch for ships . . .
Morning:	London: British Admiralty receives news that two large German ships have passed through the Kattegat on a northward course on May 20 (see May 20, 1:00 P.M.).
9:14 A.M.:	PRINZ EUGEN passes through the Korsfjord and anchors with the destroyers north of Bergen (fueled in Kalvenes Bay)—BISMARCK anchors in the Grimstadfjord.
11:00 A.M.:	Coastal Command "Spitfire" planes take off on reconnaissance flights.
1:15 P.M.:	Coastal Command "Spitfire" (Observer First Lieutenant Suckling) sights and photographs the group of

85

	BISMARCK, PRINZ EUGEN and the destroyers south of Bergen.
2:15 P.M.:	The "Spitfire" lands at Wick Airfield in northern Scotland.
3:30 & 4:00 P.M.:	The Commander-in-chief (Admiral Tovey) receives news of the successful "Spitfire" sighting.
8:00 P.M.:	BISMARCK and PRINZ EUGEN run undiscovered, along with the three destroyers, northward from Kalvenes Bay at cruising speed.
11:00 P.M.:	BISMARCK and PRINZ EUGEN leave the group (dropping the pilot) under Me 110 air protection.

Thursday, May 22, 1941:

Midnight:	Vice-Admiral Holland puts out to sea from Scapa Flow with HOOD, PRINCE OF WALES, destroyers ACHATES, ANTELOPE, ANTHONY, ECHO, ELECTRA, ICARUS.
12:15 A.M.:	The three German destroyers are released at the latitude of Drontheim.
About 12:00 noon:	BISMARCK and PRINZ EUGEN—cruising speed 24 knots—position about 65 degrees 53 minutes north, 03 degrees, 38 minutes east—Raeder reports to Hitler at the Berghof.
About 1:00 P.M.:	Hazy, rainy weather (limited visibility) with soft south winds.

Friday, May 23, 1941:

12:00 noon:	Position approximately 67 degrees 28 minutes north, 19 degrees 28 minutes west.
7:22 P.M.:	SUFFOLK reports BISMARCK and PRINZ EUGEN in the Denmark Straits and makes contact with NORFOLK . . .
About 8:30 P.M.:	BISMARCK sights NORFOLK in the fog and fires five heavy artillery salvos—the forward machine gun is knocked out—"Change numbers!" ("Nanni-Willi"): PRINZ EUGEN passes BISMARCK and takes the lead (erroneous report on order of ships on May 24, 1941, by Vice-Admiral Holland in connection with the opening of fire).
As of 11:50 P.M.:	BISMARCK and PRINZ EUGEN get into a snowstorm.

Armor Plate

The technical competition between armor plate and weapons fire had not yet been decided clearly in favor of either of the two qualities at the time when the BISMARCK was planned (1933-1934), as long as sufficient armor was provided to balance the relationship to weapons fire. By dividing the priorities among armor, firepower and speed (including range), the tactical form and the type of the warship (except for the aircraft carrier) were determined.

The armor material then used in shipbuilding had a tensile strength of up to 50 kp per square mm. The new armor materials "Wotan weich" (Ww) and "Wotan hart" (Wh), with tensile strengths of 65 to 75 kp/mm (raised 2 ' squared) and 85 to 95 kp/mm respectively, were developed for the BISMARCK. Other materials were:

—Krupp Cementite (KC): 100-mm thickness of this material corresponded in strength to 80 mm of Ww and 60 mm of Wh.
—St 52 with a strength of 52 to 64 kp/mm; this was likewise a newly developed construction steel from the firm of Gutehoffnungshütte (GHH), dating from 1929.
—St 34 and St 42, with tensile strengths of 34 to 42 and 42 to 50 kp/mm.

Material with higher tensile strength disperses the kinetic energy of the shell impact better than material of inferior quality. Thus with this material the armor protection of the BISMARCK can be improved considerably while the weight remains the same.

Regardless of this technological progress, armor had been given priority over firepower and speed in German battleship construction just as it had before World War I. This can be measured in fighting-value weights. The relationship between armor, armament and power, including auxiliary power, amounts to 40:17:9, while the corresponding values of the HOOD are 30:12:12 (rounded off in percentages).

For technical and military reasons, the armor plate cannot be qualitatively and quantitatively equally divided all over the ship. Then as before, the starting point was maintenance of the ship's fighting value, and this applies to all types of warships. What determines the fighting value of a warship is predetermined by its intended use.

In the BISMARCK—and similarly in other battleships—the following priorities applied:

a) The fighting value of the BISMARCK essentially consists of the firepower of the heavy artillery (SA) in the double turrets, and thus these particular requirements are set for their armor protection:

Front wall 360 mm; back wall 320 mm; side walls 150-200 mm; roofs 150-180 mm; floors 50-150 mm and barbettes 220 below and 340 mm above the upper deck—made of KC.

b) The ship is commanded in battle from the forward fighting command post in connection with the central position, which has been built under the armor deck with connection to the post via an aromred shaft. Here the BISMARCK is armored as follows:

Side walls 350 mm; roof 200 mm; floor 60 mm and shaft to central position 220 mm—made of KC.

c) The ship's sides are also particularly armored in the area where they are to protect powerplants, other power sources and ammunition chambers.

The belt armor (KC) has a length of some 170 meters and protects the ship's hull at the level of the armor deck with a thickness of 145 mm. The greatest thickness, about at the level of the waterline, measures 320 mm and tapers off to 170 mm at the bottom of the ship. The citadel armor (KC) above the belt armor, which reaches to the edge of the upper deck and is anout the same length, has a thickness of 100 mm.

The side armor is attached to a 50-mm layer of teakwood.

Additional side protection is provided by a longitudinal torpedo shaft of 45-mm thickness.

Transverse armored shafts forward of turret A, with thicknesses of 145 to 220 mm, and aft of turret D, with thicknesses of 110 to 220 mm, form the boundaries of the side armor.

d) Under the upper deck, which is protected by 50-mm armor (Wh) throughout, an armor deck (Wh) is built, chiefly for protection from air attack; it inclines downward toward the ship's sides and is offset to the ship's bow and stern. This armor, beginning at about 40 mm before the point of the bow, has a thickness of 80 mm amidships and 100 mm in the turret areas. This armor is thickest aft, measuring 110 mm, in order to protect the rudder apparatus from attack from above.

British ships such as the HOOD—presumably proved by its explosion—are said to be insufficiently protected. This applies only in that the armor deck was not covered along its center line at about the width of the funnel. Another weak point in its self-defense was the insufficient division of the ammunition and powder magazines. In other areas the HOOD was strongly armored. Along the waterline, for example, the armor thicknesses added up to more than 400 mm.

(in mm)	Decks	Front and Side	Floor
15-mm turrets	20-35	40-100 (KC)	20
Aft command post	50	150	30
Artillery post	20	60	20

The unarmored parts of the ship were still protected against the shrapnel effect of heavy artillery shells.

During the BISMARCK's final battle, the British fired 2876 shells at the German battleship.

Of them there were:

380 of 41-cm caliber,
339 of 35.6-cm caliber,
781 of 20.3-cm caliber,
716 of 15.2-cm caliber, and
66 of 13.3-cm caliber.

There are no definite statistics on how many of them hit. Because of the inability to maneuver and the sometimes short distance, it can nevertheless be assumed that a large percentage of the shells hit, which is also attested to by the survivors. In addition, there were 8 definitely registered torpedo hits. It is known that the Bismarck remained capable of staying afloat. And as eyewitnesses report, the hull was not penetrated anywhere in the belt armor. Thus according to constructive points of view, the German battleship was quasi-unsinkable.

But the BISMARCK operation in itself shows that this was no criterion for the fighting value of this ship. It showed in the end that the BISMARCK was worth as much as its most vulnerable spots, for-

—the two hits that the BISMARCK took from the HOOD, without markedly decreasing its fighting value, were nevertheless so serious that the operation could no longer be carried out in its intended form, and

—the torpedo hit in the port rudder system, which did not decrease the BISMARCK's firepower, made it useless as a weapon of war.

There is no point in calculating the chances of this tragic torpedo hit. It should be seen from the start that the hit or hits in the ammunition chamber of the HOOD, which led to its sudden destruction, were just as much as matter of chance and just as tragic in their results.

Even heavy armor does not always offer the expected protection. It can prevent penetration, of course, but not the secondary effects that can lead to lasting damage and the failure of the apparatus. Just 17 minutes after fire on the BISMARCK commenced, both of its two heavy forward turrets were put out of commission by a hit from the RODNEY, and almost all the bridge personnel were killed in the process.

The principle of either "over- or underarmoring" was not only a quality of the BISMARCK, but was likewise introduced in the modernized and new major warships on the British Navy. Thus the belt armor of the HOOD had a thickness of about 300 mm, that of the RODNEY almost 360 mm; yet the BISMARCK, as already noted above, had an advantage via the use of better armor materials.

And finally, "chance" determined the events of Operation "Rhein-übung", and no ship can be so heavily armored that the unexpected is ruled out.

To give an impression of the mighty strength (armor) of the BISMARCK, it is not necessary to present a detailed description of the armor plate. The massive front view of the BISMARCK in drydock (left and far right) is much more impressive.

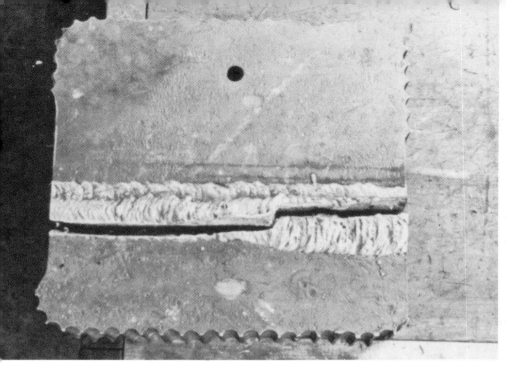

The new armor steels that were used place high demands on the technique of welding. Lasting firmness, especially in the highly stressed parts of the ship, require constant checking.

Here a piece of two welded armor plates has been cut out. The welded seam has been opened longitudinally and is now ready for metallurgical testing.

Next page:
The forward anti-aircraft calculating post is set up. As in this room, even under peacetime conditions most of the crewmen put in their service behind armor protection and without natural light.

The foundation of the side armor is 50-mm blocks of teakwood. To these the armor plates are mounted (light armor plates) by being screwed onto the wood blocks and welded together. In the belt-armor area additional armor plates (dark color) are welded first to the side armor and then to each other.

Previous page:
The hull of the BISMARCK consists of an "armored box" some 180 meters long, which is closed off by armored transverse bulkheads about 40 meters aft of the bow and 30 meters forward of the stern. The angle of the belt armor, which sets the longitudinal boundaries of the "armored box" and reaches maximum thicknesses of 320 mm, clearly differs from that of the hull. The width amounts to three meters over and two meters under the waterline.

In almost exactly fifteen months, a British torpedo plane would strike this part of the BISMARCK under the light gray area with a torpedo. Its ability to fight and stay afloat would not be weakened in the process.

Yet this blow was so fatal that the BISMARCK would be exposed, almost defenseless, to the fire of the attacking Royal Navy ships.

The German Navy would lose at one stroke more than 2000 officers and men, as well as irreplaceable war materials that took up a large part of the German shipyard capacity for years.

Next page:
The greatest armor thicknesses and weights were concentrated in the area of the heavy artillery turrets as well as under the deck. The heavy turrets weighed about 4000 tons in total.

Flottenchef Admiral LÜTJENS

The Men Involved in Operation "Rheinübung"

Fleet Chief Admiral Günther Lütjens

5/25/1889:	born in Wiesbaden
4/3/1907:	entered Imperial Navy, crew of 1907, appointed sea cadet
4/3/1907-4/8/1907:	Naval School
5/9/1907-3/31/1908:	Heavy Cruiser FREYA (training ship)—basic seaman training during a Mediterranean cruise—inspectorate of naval training system
4/1/1908-3/31/1909:	Naval School, Kiel
4/21/1908:	Promotion to Ensign
4/1/1909-6/30/1909:	Naval Artillery School
7/1/1909-8/31/1909:	Torpedo training and testing ship WURTTEMBERG
9/1/1909-9/30/1909:	Second Sea Battalion
10/1/1909-9/14/1910:	Battleship ELSASS
9/15/1910-9/25/1910:	At command of First Naval Inspection
9/26/1910-3/31/1911:	Mürwik: naval apprentice trainer
9/28/1910:	Promotion to Lieutenant
4/1/1911-3/31/1913:	Naval cadet teacher—naval cadet training ship HANSA—inspectorate of naval training
4/1/1913-9/30/1913:	Mürwik: naval apprentice trainer
9/27/1913:	Promotion to First Lieutenant
10/1/1913-10/31/1913:	First Torpedo Division
11/1/1913-12/21/1913:	Third Torpedo Boat Flotilla—Sixth Torpedo Semi-Flotilla on G-169—simultaneously company officer: Sixth T-Reserve Semi-Flotilla
12/24/1913-3/14/1914:	Company officer: First Torpedo Division—at times commandant of a torpedo boat
3/15/1914-7/31/1914:	Inspectorate of Torpedo Operations in Kiel —Sixth Torpedo Semi-Flotilla-simultaneously company officer: Sixth T-Reserve Semi-Flotilla
8/1/1914-12/6/1914:	Harbor flotilla—Jade—as of 9/4/1914 Commander of T-68 (ex-S-68)
12/7/1914-1/1/1915:	First Torpedo Division
1/2/1915-1/15/1915:	Minesweeping training
1/16/1915-3/13/1915:	First Torpedo Division: training boat commander
3/14/1915-5/4/1915:	First Torpedo Division
5/5/1915-11/10/1918:	Torpedo boat flotilla—Flanders—Torpedo boat commander: A-5 and A-20—as of February 1916 Chief

	of the Second A-Semi-Flotilla—Flanders—simultaneously Commander of A-40
...as of 6/16/1915:	Regular night patrol service off Zeebrugge with A-5 (Lütjens)
8/5/1915:	A-5 (Lütjens)—advance with A-7 and A-8 to Thornton Bank (salvage of a French seaplane)
8/22/1915-8/23/1915:	A-5 (Lütjens)—advance with A-13 and A-14 to mouth of Scheide (action against merchant shipping)
11/28/1915:	A-5 (Zeebrugge guard boat) tows in a downed airplane with air support
5/25/1916:	Torpedo flotilla—Flanders: A-boat mine reconnaissance
3/25/1917, 11:15 PM:	From Ostende: A-39, A-40, A-42 (Lütjens) & A-45—artillery attack on Dunkerque from 2:16 to 2:22 AM on 3/26/1917 (return date)
4/24/1917, 9:30 PM:	From Zeebrugge to night operations with A-39, A-40, A-42 (Lütjens) & A-45—to the Stroombank buoy off Ostende, combat with two S-boats—2:15-2:22 AM (April 25): artillery attack on Dunkerque—retreating action with the French T-boat ETENDARD—return: 4/25/1917
5/2/1917, 9:25 PM:	Search operation (for a downed airplane) with A-42 (Lütjens) and A-40—10:45 PM: artillery action against British S-boats north of Middelkerke Bank—return: 5/3/1917
5/19/1917, 9:45 PM:	Under leadership of Group Leader (Lütjens) operation of Second Torpedo Semi-Flotilla in direction of eastern exit of Zuidcoot Pass: A-42 (Lütjens), A-39, A-40, A-43 & A-45—action with French T-boats—the units BOUCLIER (8 dead, 11 wounded) and CAPITAINE MEHL were hit—return: 5/20/1917
5/24/1917:	Promotion to Lieutenant Commander
7/12/1917-7/13/1917:	Joint action with three C-planes against merchant shipping: prizes brought in by A-40 (Lütjens) and A-46
7/25/1917:	A-boats of the Second Semi-Flotilla: action against British barrage group
8/5/1917-8/6/1917:	A-40 (Lütjens) brings in A-42, damaged by a mine
9/24/1917:	A-40 (Lütjens) and A-50 as well as V-71 and V-73 in supporting action north of Middelkerke Bank—action against five British destroyers and five enemy planes
9/28/1917:	Brought in the submarine U-70
11/15/1917:	Securing activity
11/17/1917:	Run to southeast of Buoy 13 in test cruise of Second Torpedo Semi-Flotilla Flanders (Lütjens) A-50 and S-54 on mines

2/17-18/1918:	A-40 badly damaged in air raid on Bruges shipyards
4/22-23/1918:	British attempt to blockade Zeebrugge and Ostende
4/24/1918:	Boats of Second Torpedo Semi-Flotilla Flanders leave Zeebrugge for routine removal operation
11/11-23/1918:	Retreat from Antwerp to Kiel (war ends)
11/15/1918:	A-40 and A-42, among others, left behind in withdrawal from bases in Flanders

Decorations

World War I: Knight's Cross with swords,
Order of the House of Hohenzollern
Iron Cross second and first class
Order of the Lion of Zähringen/Knight's Cross
Hamburg Hanseatic Cross
Oldenburg Friedrich August Cross, First Class

11/24/1918-11/30/1918:	Kiel: under command of Baltic Sea Naval Station
12/1/1918-1/23/1919:	Warnemünde: Leader of sea transport unit
1/24/1919-2/7/1919:	Lübeck: Leader of sea transport unit
2/8/1919-3/9/1919:	Warnemünde: Leader of sea transport unit
3/10/1919-7/7/1919:	Admiralty duty
7/8/1919-9/14/1919:	Lübeck: Leader of sea transport unit
9/15/1919-6/6/1921:	Company leader: Coast guard unit III/Camp IV
6/7/1921-9/30/1923:	Berlin: Administrator in naval command: Naval Command Office A, Fleet Section A II (Gladisch/Assmann, Lütjens/Lindemann), 1922: General Navy Office B
10/1/1923-9/25/1925:	Second Torpedo Flotilla: Chief of Third Torpedo Semi-Flotilla
9/26/1925-10/2/1929:	Wilhelmshaven: North Sea Naval Station, First Adjutant and Personnel Officer
4/1/1926:	Promotion to Corvette Captain
8/1/1926-8/31/1926:	Supernumerary assignment to naval yacht ASTA
12/5/1927-12/9/1927:	Staff officer-torpedo training
4/21/1928-4/28/1928:	Supernumerary assignment to battleship SCHLESIEN
8/14/1928-8/18/1928:	Supernumerary assignment to battleship SCHLESIEN
10/3/1929-1/8/1930:	Swinemünde: Chief of the First Torpedo Boat Flotilla
1/9/1930-1/11/1930:	Staff officer training for leadership position
1/12/1930-2/2/1930:	Chief of the First Torpedo Flotilla
2/3/1930-2/7/1930:	Torpedo and Intelligence School: torpedo training for staff officers
2/8/1930-2/1/1931:	Chief of the First Torpedo Flotilla

2/2/1931-2/6/1931:	Torpedo and Intelligence School: training for commanders and staff officers in leadership positions
2/7/1931-2/15/1931:	Chief of the First Torpedo Flotilla
2/16/1931-2/21/1931:	Berlin: Naval Command: training for technical navigation
2/22/1931-9/16/1931:	Chief of the First Torpedo Flotilla
9/17/1931-9/25/1931:	Berlin: Ministry of Defense/Naval Command: Administrator A IIa, 1931: Lütjens, as A IIa in Fleet Section (Boehm), plans and organizes the first fleet maneuver with aircraft, 1932: Naval Command: Department Director of the Naval Officer Personnel Department (MPA)
10/1/1931:	Promotion to Frigate Captain
9/26/1932-9/15/1934:	Berlin: Ministry of Defense: Director of the Naval Personnel Office
7/1/1933:	Promotion to Captain
8/20/1933-8/22/1933:	Battleship HESSEN: supernumerary assignmentfor fleet torpedo firing
9/16/1934-9/23/1935:	Commander of the Light (training) Cruiser KARLSRUHE
10/22/1934:	Light Cruiser KARLSRUHE (Captain Lütjens) leaves Kiel with part of the naval officer class of 1934 (Crew 34) for a foreign (training) cruise of several months
6/1/1935-6/8/1935:	Vigo, Spain—meeting with Cruiser EMDEN, return trip home together
6/12/1935:	KARLSRUHE and EMDEN arrive at Schillig Roadstead, Wilhelmshaven, both commanders report to ObdM (Raeder)
6/15/1935:	KARLSRUHE arrives at Kiel (Naval Week)
9/24/1935-3/15/1936:	Wilhelmshaven: North Sea Naval Station, Chief of Staff
3/16/1936-10/7/1937:	Berlin W-35, Tirpitz-Ufer 72-76: Chief of Naval Personnel Office (MPA) in Supreme Command of the Navy (OKM)
10/1/1937:	Promotion to Rear Admiral
10/8/1937-10/20/1939:	Commander of the Torpedo Boats (FdT): Destroyers, torpedo boats and high-speed boats, command ship: Destroyer LEBERECHT MAASS
11/9-10/1938:	"Crystal Night" (destruction of Jewish stores, action against Jews), Lütjens protests to ObdM (Raeder)
9/1/1939:	Beginning of German attack on Poland: outbreak of World War II
10/17/1939-10/18/1939:	Offensive FdT mine operation with six destroyers in the North Sea, success: 7 craft—25,825 tons
1/1/1940:	Promotion to Vice-Admiral
3/4/1940-4/13/1940:	Fleet Chief i.V.
	as of 4/7/1940: Operation "Weserübung" (Scandinavian operation, Norway and Denmark): cover group for Narvik and Trondheim under command of Vice-Admiral Lütjens with battle cruisers GNEISENAU and SCHARNHORST
4/9/1940:	Vestfjord: short battle between GNEISENAU and SCHARNHORST and British battle cruiser RENOWN
6/14/1940:	Conferring of the Knight's Cross to the iron Cross
6/18/1940-7/8/1940:	Fleet Chief i.V.
9/1/1940:	Promotion to Admiral
12/28/1940-1/2/1941:	Operation against merchant shipping by the battle cruisers GNEISENAU and SCHARNHORST must be broken off
1/22/1941-3/23/1941:	Operation "Berlin" with battle cruisers GNEISENAU and SCHARNHORST, action against merchant shipping in the Atlantic, success: 22 units sunk, total 115,622 tons
5/19/1941-5/27/1941:	Operation "Rheinübung" with battleship/ fleet flagship BISMARCK (Captain Lindermann) and heavy cruiser PRINZ EUGEN (Captain Brinkmann)
6/1/1941:	PRINZ EUGEN reaches Brest
8/11/1967:	Bath, USA: Bath Iron Works, christening of the first West German guided-missile destroyer with the name LUTJENS

Grand Admiral Erich Raeder, Ph.D., born 4/24/1876 in Wandsbek, near Hamburg, died 11/6/1960 in Kiel—entered the navy 4/16/1894 (Crew 94)—naval cadet: 5/13/1895—Second Lieutenant: 10/25/1897--First Lieutenant: 4/9/1900—Lieutenant Commander 3/21/1905—Corvette Captain: 4/15/1911—Frigate Captain: 4/26/1917—Captain: 11/29/1919-Rear Admiral: 8/1/1922—Vice-Admiral: 4/1/1925-Admiral: 10/1/1928—Fleet Admiral: 4/20/1936-Grand Admiral: 4/1/1939—Commands: STOSCH, GNEISENAU, MARS, BLUCHER, SACHSEN, BADEN, DEUTSCHLAND, AEGIR, GRILLE,

KAISER WILHELM DER GROSSE, KAISER FRIEDRICH III, FRITHJOF, YORK, HILDE-BRAND, HOHENZOLLERN, MOLTKE, SEYD-LITZ, COLN, HAMBURG, ELSASS, SCHLES-WIG-HOLSTEIN—May-June 1916: Participation in the battle of the Skaggerak on board SEYDLITZ—10/1/1928 to 5/31/1935: Chief of naval command—6/1/1935 to 1/31/1943: Commander of the Navy (ObdM)—2/1/1943 to 5/8/1943: Inspector Admirak of the Navy—1945-46: International Military Tribunal in Nünberg: 1946-55: Sentence (war criminal) and confinement at Spandau.

Admiral of the Fleet Lord John Cronyn Tovey, of Langton Maltravers, G.C.B. (K.C.B., C.B.); K.B.E., D.S.O.—born 3/7/1885, died 1/11/1971—Captain: 12/31/1923—Rear Admiral 8/27/1935—Vice-Admiral: 5/3/1939—1915-18: Destroyer commander (Skaggerak battle aboard ONSLOW)—other commands include: FAULKNOR, BRITANNIA, MAJESTIC, AMPHION, RODNEY—1930-32: Naval Assistant to Second Sea Lord—1932-34: RODNEY—1935: Flag officer, at royal court—1938-40: Rear Admiral, Destroyers in the Mediterranean—June 1940: Mediterranean, Chief 7th Cruiser Squadron—6/20-21/1940: Mediterranean, British-French operations against Bardia--Securing the MA-3 convoys—July 1940: Mediterranean, Second-in-command, Commander, Force A—12/2/1940-5/8/1943: Commander-in-Chief, Home Fleet-

January 1941: Home Fleet operation against SCHARNHORST and GNEISENAU (Operation "Berlin")—May 1941: action against BISMARCK on battleship KING GEORGE V (photo)-November 1941: Home Fleet, US battle groups—Action against suspected TIRPITZ outbreak (from Norway) into the Atlantic—March 1942: Polar Sea PQ-12 operation—June-July 1942: Polar Sea QP-13, PQ-17 operations—1943-1946: Commander-in-Chief, The Norre—Admiral of the Fleet—1945-46: First and Principal Naval at royal court—1948-52: Third Estates Commissioner.

Admiral Sir James Somerville—July 1940: Mediterranean Operation "Catapult"—Naval battle at Punta Stilo, Calabria (including Force H, with HOOD etc.)—November 1940: Mediterranean operation "White" (Force H)—Operation "Collar" (naval battle at Cap Teulada, Sardinia)—January 1941: Mediterranean operation "Excess" (reinforcing Malta)—January-February 1941: Mediterranean operation against Sardinia and Henoa—May 1941: Action against BISMARCK—Gibraltar Squadron, Force H with the flagship RENOWN, aircraft carrier ARK ROYAL and cruiser SHEFFIELD—May 1941: Mediterranean operation "Splice"-September 1941: Mediterranean operation "Halberd" (supplying Malta)—October-November 1941: Force H operation—Carrier ARK ROYAL (U-331/Tiesenhausen)—April 1942: Indian Ocean—Japanese Ceylon operation—Action of the British Eastern Fleet—January-February 1944: the British Eastern Fleet is strengthened—March-April 1944: Operation Cocos Islands, Trincomalee, Ceylon—April 1944: Operation "Cockpit"—May 1944: Operation "Transom"—July 1944: Operation "Crimson"—8/8/1944: Change in leadership of the British Eastern Fleet—Admiral Fraser replaces Admiral Somerville.

Admiral (Third Sea Lord, Controller of the Navy) Sir William Frederic Wake-Walker, K.C.B., C.B., C.B.E., O.B.E.—born 3/24/1888, died 9/24/1945—1904: Naval cadet—June 1920: Commander-12/31/1927: Captain—1/10/1939: Rear Admiral—1942: Vice-Admiral—1945: Admiral—Commands include GOOD HOPE, INVINCIBLE, COCHRANE, RAMILLIES, ROYAL OAK—1928-30: Commander of CASTOR—1932-35: Commander of DRAGON—1938: Commander of REVENGE--Until 1925: Royal Navy Staff College, Greenwich, Operations and Tactical Divisions of the Admiralty Naval Staff—until 1932: Admiralty, Deputy Director of the Training and Staff Duties Division of the Naval Staff—1935-37: Admiralty, Torpedo and Mine Section—1939: Flag Officer rank—May-June 1940: Channel, Dunkerque operation "Dynamo"—loss of the flagship-flotilla leader KEITH—1940-41: Cruiser Squadron Chief—May 1941: action against BISMARCK (flagship NORFOLK)—August-September 1941: Polar Sea service—among others, securing of the first experimental escort convoy "Dervish"—1942-45: Controller of the Navy—9/2/1945: Pacific: aboard the battleship NELSON, accepted the Japanese surrender of the Penang area.

Lieutenant Commander Herbert Wohlfahrt ("Parzival")—born 6/5/1915 in Kanagawa, Japan—Entered the Navy in 1933—10/1/1940: Lieutenant Commander—1939-40: Commander of U-14 (sank 9 craft, 12,362 tons)—1940: Commander of U-137 (sank 6 craft, 19,557 tons)—1940-41: Construction training U-556 (Blohm & Voss boat)—2/6-6/27/1941: Commander of U-556 (sank 5 craft, 23,557 tons)—Knight's Cross of the Iron Cross (5/15/1941), conferred personally by BdU (Döitz) in May 1941 after return from the first U-556 operation (BISMARCK support)—Total sinkings with three U-boats: 20 craft, 55,476 tons—Prisoner of war: July 1941 to July 1947.

Captain John Catterall Leach aboard the battleship **PRINCE OF WALES**—August 8-12, 1941: Argentia Bay, Newfoundland—meeting of Churchill and Roosevelt ("Atlantic Charter") on US cruiser **AUGUSTA**, battleship **PRINCE OF WALES**—On 12/10/1941, the **PRINCE OF WALES** under Captain Leach, with Admiral Sir T. S. V. Phillips on board, was sunk by torpedo planes and bombers, along with the battle cruiser **REPULSE** (Tennant), east of Malaya after a splendid attack on the 22nd Japanese Naval Squadron commanded by Rear Admiral Matsunaga (327 dead)—the battleship had a bad reputation in the Royal Navy after the fight with the **BISMARCK** (disengaging action of the badly hit but still operational ship).

Vice-Admiral Lancelot Ernest Holland (1887-1941), C.B.—Artillery specialist—Captain: 6/30/1926-Rear Admiral: 1/11/1938—1930-31: Chief of the Naval Mission for Greece (Commander of the Order of the Redeemer)—1937-38: Assistant Chief of Naval Staff—1938-39: 2nd Battle Squadron—1939-40: Admiralty, Air Ministry Staff—April 1940: Norwegian action on **RODNEY**—November 1940: Mediterranean operation "Collar", naval battle at Cap Teulada, Spartivento, Sardinia—1941: Commanding the Battle Cruiser Squadron, Second-in-Command of the Home Fleet (**HOOD**).

Captain Ralph Kerr, C.B.E., R.N.—Captain: 6/30/1935—1940: Commander of the HOOD.

Fleet Chief Admiral Lüjens (left) aboard the flagship GNEISENAU. In the center is the Chief of Staff, Captain Harald Netzband (died on the BISMARCK): joined the Navy in 1912—10/1/1937: Captain—1935-39: OKM/MPA Department Chief—1939-40: Commander of GNEISENAU—1940-41: Fleet Command Chief of Staff. Right: Fleet Engineer Frigate Captain (Eng.) Engineer Karl Thannemann (died on the BISMARCK): joined the Navy in 1918—2/1/1940: Frigate Captain (Eng.), posthumously promoted to Captain (RDA 5/1/1941)—1939-40: Chief Engineer (LI) of BLUCHER--1940: under command of the Baltic Sea Station Command, Kiel—OKM/K IV (Technical Advisor)—1940-41: Fleet Command Fleet Engineer.

The BISMARCK's commander, Captain Ernst Lindemann, photographed in Hamburg (Blohm & Voss shipyards) on the day of the BISMARCK's commissioning, 8/24/1940 (crewmate of the PRINZ EUGEN's Commander Brinkmann), born 3/28-1894 in Altenkirchen, Rheinland, died 5/27/1941 aboard the battleship BISMARCK at approximately 48 degrees 10 minutes North, 16 degrees 12 minutes West.

Participating British Fleet Units

Battleship	KING George V	Captain Wilfrid Rupert Patterson C.I.C. Flagship: Admiral Sir John Cronyn Tovey
	PRINCE OF WALES	Captain John Catterall Leach
	RAMILLIES	Captain Arthur Duncan Read
	REVENGE	Captain Ernest Russell Archer
	RODNEY	Captain Frederick Hew George Dalrymple-Hamilton
Battle Cruiser	HOOD	Captain Ralph Kerr Flagship Group Chief Vice Admiral Lancelot Ernest Holland
	RENOWN	Captain Rhoderick Robert McGrigor Flagship Force H (Gibraltar) Vice-Admiral Sir James Fownes Somerville
	REPULSE	Captain William George Tennant
Aircraft Carrier	ARK ROYAL	Captain Loben Edward Harold Maund
	VICTORIOUS	Captain Henry Cecil Bovell
Heavy Cruiser	DORSETSHIRE	Captain Benjamin Charles Stanley Martin
	LONDON	Captain R. M. Servais
	NORFOLK	Captain Alfred Jerome Lucian Philipps Flagship (1st Cruiser Group), Rear Admiral William Frederick Wake-Walker
	SUFFOLK	Captain Robert Meyrick Ellis
Light Cruiser	ARETHUSA	Captain Lex Colin Chapman
	AURORA	Captain William Gladstone Askew
	BIRMINGHAM	Captain Alexander Cumming Gordon Madden
	EDINBURGH	Captain Charles Maurice Blackman
	GALATEA	Captain Edward William Boyd Sim Flagship (2nd Cruiser Group), Rear Admiral Alban Thomas Buckley Curteis
	HERMIONE	Captain Geoffrey Nigel Oliver
	KENYA	Captain Michael Maynard Denny
	MANCHESTER	Captain Herbert Annesley Packer
	NEPTUNE	Captain Rory Chambers O'Connor
	SHEFFIELD	Captain Charles Arthur Aiskew Larcom
Destroyer	ACHATES	Lieutenant Commander Viscount Jocelyn
	ACTIVE	Lieutenant Commander Michael Wilfred Tomkinson
	ANTELOPE	Lieutenant Commander R. B. N. Hicks
	ANTHONY	Lieutenant Commander J. M. Hodges
	ASSINIBOINE	Commander G. C. Jones
	COLUMBIA	Lieutenant Commander S. W. Davis
	COSSACK	Captain Philip Louis Vian (also Chief 4th Destroyer Flotilla, Command Ship COSSACK)
	ECHO	Lieutenant Commander C. H. de B. Newby
	ELECTRA	Commander Cecil Wakeford May
	ESKIMO	Lieutenant J. V. Wilkinson
	FAULKNOR	Captain Antony Fane de Salis
	FORESIGHT	Commander Jocelyn Stuart Cambridge Salter
	FORESTER	Lieutenant Commander Edward Bernard Tancock
	FOXHOUND	Commander Geoffrey Hendley Peters
	FURY	Lieutenant Commander Terence Corin Robinson
	HESPERUS	Lieutenant Commander Arthur Andre/ Tait
	ICARUS	Lieutenant Commander Colin Douglas Maud
	INGLEFIELD	Captain Percy Todd
	INTREPID	Commander Roderick Cosmo Gordon
	JUPITER	Lieutenant Commander Norman Vivian Joseph Thompson Thew
	LANCE	Lieutenant Commander Ralph William Frank Northcott
	LEGION	Commander Richard Frederick Jessel
	MAORI	Commander Harold Thomas Armstrong
	MASHONA	Commander William Halford Selby
	NESTOR	Commander Conrad Byron Alers-Hankey
	PUNJABI	Commander Stuart Austin Buss
	PIORUN (Polish)	Commander E. Plawski
	SAGUENAY	Commander G. R. Miles
	SIKH	Commander Graham Henry Stokes
	SOMALI	Captain Clifford Caslon
	TARTAR	Commander L. P. Skipwith
	WINDSOR	Lieutenant Commander the Hon. John Montagu Granville Waldegrave
	ZULU	Commander Harry Robert Graham
Submarine	H-44	Lieutenant W. N. R. Knox
	MINERVE (Fr.)	Lieutenant de Vaisseau P. M. Sommerville
	PANDORA	Lieutenant Commander John Wallace Linton
	P-31	Lieutenant J. B. de B. Kershaw
	SEALION	Commander B. Bryant
	SEAWOLF	Lieutenant P. L. Field
	STURGEON	Lieutenant Commander Drummond St. Clair Ford
	TIGRIS	Lieutenant Commander H. F. Bone

Aerial view of the battle cruiser HOOD, the largest warship in the world from 1921 to 1941.

Vice-Admiral Sir W. Frederick Wake-Walker. As Rear Admiral, Commander of the Cruiser Group NORFOLK (Flagship) and SUFFOLK.—Heavy cruiser H.M.S. NORFOLK—Displacement: 9975 tons—Built: 1930—Main armament: 8 20.3-cm guns—Commander in May 1941: Captain Alfred Jerome Lucian Phillips—Likewise flagship of Rear Admiral Commanding, 1st Cruiser Squadron: Rear Admiral William Frederick Wake-Walker—This vessel was engaged in the pursuit of the BISMARCK longest (as of 5/23/1941, 8:30 P.M., BISMARCK fired 355 heavy artillery salvos on NORFOLK)—In the last phase of the battle, the cruiser fired on BISMARCK on May 27 between 9:45 A.M. (NORFOLK opened fire) and 10:15 A.M. (Commander-in-Chief halted fire), firing 527 20.3-cm shells and eight torpedoes (below).

Heavy Cruiser H.M.S. SUFFOLK—Displacement: 9800 tons—Built: 1928—Main armament: 8 20.3-cm guns—Commander in May 1941: Captain Robert Meyrick Ellis—The photo shows the cruiser (with camouflage paint, May 1941) in the Denmark Straits. It reported the BISMARCK-PRINZ EUGEN group on 5/23/1941 at 7:22 P.M.: One battleship, one cruiser in sight at 020 degrees—distance seven miles—course 240 degrees...(above).

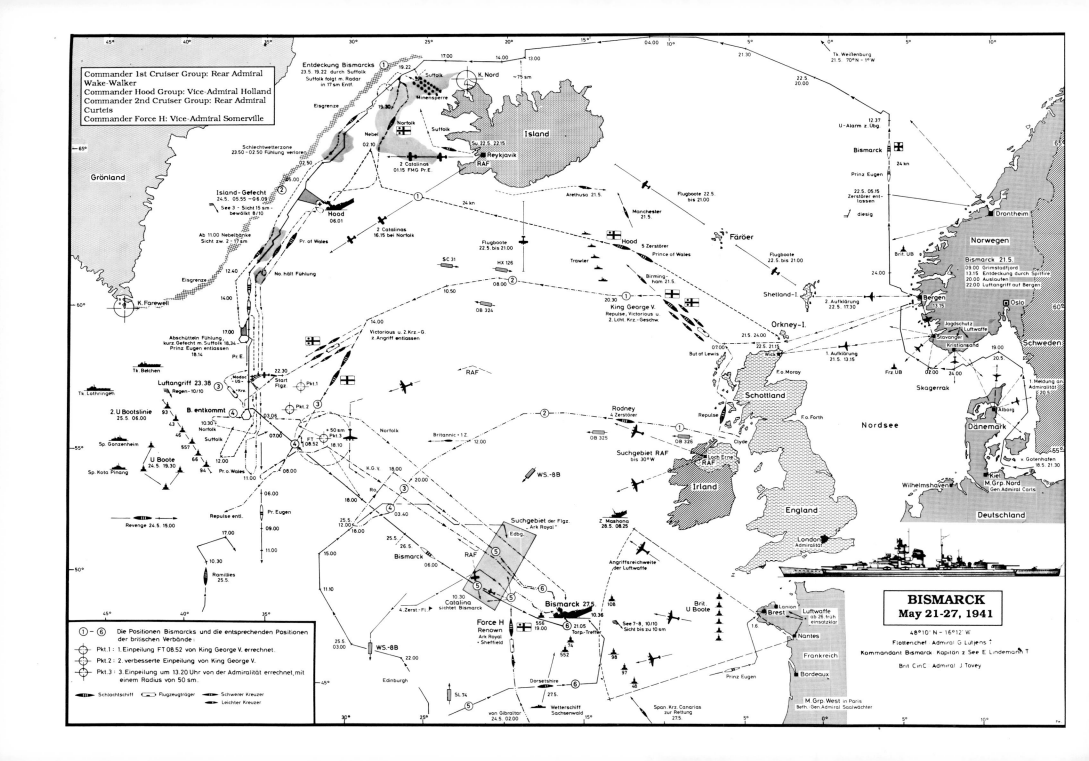

BISMARCK
May 21-27, 1941

48°10' N – 16°12' W

Flottenchef Admiral G. Lütjens †

Kommandant Bismarck Kapitän z. See E. Lindemann †

Brit. CinC Admiral J. Tovey

Commander 1st Cruiser Group: Rear Admiral Wake-Walker
Commander Hood Group: Vice-Admiral Holland
Commander 2nd Cruiser Group: Rear Admiral Curteis
Commander Force H: Vice-Admiral Somerville

① – ⑥ Die Positionen Bismarcks und die entsprechenden Positionen der britischen Verbände:

⊕ Pkt.1 : 1.Einpeilung FT 08.52 von King George V. errechnet.
⊕ Pkt.2 : 2. verbesserte Einpeilung von King George V.
⊕ Pkt.3 : 3.Einpeilung um 13.20 Uhr von der Admiralität errechnet, mit einem Radius von 50 sm.

▬▬ Schlachtschiff   ▭ Flugzeugträger   ◥ Schwerer Kreuzer
◢ Leichter Kreuzer

M. Grp. West in Paris
Beth. Gen. Admiral Saalwächter

The Iceland Sea Battle

The Ships Directly Involved

Battle Cruiser H.M.S. HOOD

Flagship of the Group Chief of the Battle Cruiser Squadron:
Vice-Admiral Lancelot Ernest Holland
Artillery Officer (AO): Lieutenant Commander Edward Home
Battleship H.M.S. PRINCE OF WALES
Commander: Captain John Catterall Leach
Artillery Officer: Lieutenant Commander Colin William McMullen
Heavy Cruiser H.M.S. NORFOLK
Commander: Captain Alfred Jerome Lucian Phillips
Flagship of the Group Chief of the 1st Cruiser Squadron:
Rear Admiral William Frederick Wake-Walker
Heavy Cruiser H.M.S. SUFFOLK
Commander: Captain Robert Meyrick Ellis
Battleship BISMARCK
Commander: Captain Ernst Lindemann
Fleet Flagship of Fleet Chief:
Admiral Günther Lütjens
Artillery Officer: Corvette Captain/Frigate Captain Adalbert Schneider
Note: all of them perished aboard BISMARCK on May 27, 1941
Heavy Cruiser PRINZ EUGEN
Commander: Captain Helmuth Brinkmann
Artillery Officer: Corvette Captain Paul Jasper

May 24, 1941: 5:53 to 6:09 A.M.

5:52-5:53 A.M.: Weather: northeast wind, strength 3-4, sea 2-3 (at times—see photos, film—wind: 1-3, sea 0-1/2) —visibility 15 nautical miles, cloud cover 8/10-10/10- almost simultaneously British fire opened (only with forward turrets: unfavorable without speed advantage coming up from aft, thus no forward, favorable position at beginning of battle) by HOOD and PRINCE OF WALES, turning somewhat to port (distance: approximately 289 hm) via optical range measurement (not radar!) against German battle group: HOOD (presumably) against the leading German ship PRINZ EUGEN (mistaken announcement of target, confusion of PRINZ EUGEN with BISMARCK) . . . PRINCE OF WALES, contrary to command of Admiral Holland, fired 5 35.6-cm guns (Turret A ' quadruple turret, Turret B ' double turret) against BISMARCK (correct announcement of target)— aboard PRINZ EUGEN the units firing are identified as major warships . . . The first HOOD salvo(s) were fired at PRINZ EUGEN—the HOOD's hits (height approximately 30-35 to 40 meters) struck 100 to 150 meters and 200 to 300 meters forward, port side (approximately 330 to 340 degrees by ship's location)—PRINZ EUGEN turned directly into the columns of water and thus toward the enemy . . .

5:35 A.M.: PRINCE OF WALES fires both its forward turrets (A and B) at BISMARCK . . .

At/after 5:53 A.M.: NORFOLK and SUFFOLK observation, from distance of approximately 12 to 15 nautical miles: 1st and 2nd HOOD salvos fell "close to the enemy" (BISMARCK or PRINZ EUGEN?), 3rd salvo "covering"—1st PRINCE OF WALES salvo (by calculation of forward 5-m artillery command post) fell far away (about 800 to 900 meters). . . 6th salvo fell "covering"—all SUFFOLK salvos fell too short . . .

ca. 5:54 A.M.: Presumed hit on BISMARCK, forward (3rd or 4th PRINCE OF WALES countersalvo?)—according to Grenfall, black funnel smoke became visible . . .

5:55 A.M.: German on-board time: Fire opened ("JOT DORA"/"JD") by BISMARCK and PRINZ EUGEN: fire concentrated on HOOD—Target announced aboard PRINZ EUGEN: at first thought to be a cruiser or destroyer—Target announced aboard BISMARCK: I.A.O. (Schneider): Cruiser . . . II.A.O. (Albrecht: HOOD . . . BISMARCK and PRINZ EUGEN return fire with a certain delay (Lütjens decision: several queries from aboard BISMARCK: clearance to fire?), fire optically (almost) simultaneously, with favorable course and full (full salvos) broadsides immediately on target (' with all turrets A-B and C-D, on BISMARCK also with port I-III 15.0-cm double turrets against PRINCE OF WALES, heavy artillery turret salvo — ca. 15"—medium artillery turret salvo ' ca. all 9")—PRINZ EUGEN: "Commander to artillery: Heavy, clearance to fire!"—One salvo!—Four turrets: 8 shells: eight short! (1st salvo (situation): 4 shots fell far)—HOOD and PRINCE OF WALES corrected course (Formation: short starboard formation) to port (almost) parallel course (equal course) to PRINZ EUGEN and BISMARCK-German salvos not observable due to group nature (per salvo ' 4 shots, A-B, C-D) fire from BISMARCK and PRINZ EUGEN: the smoke blowing southward reaches the enemy, making German target observation difficult; for British observation, the powder smoke moved off quite rapidly in the opposing wind (good observation possibility)—Fire against HOOD immediately fell covering with very little scattering: "The artillery superiority of the enemy was equalized in that he could not bring his after turrets into the battle on account of his angle of approach."

ca. 5:55 A.M.: HOOD has correct range . . .
5:55-6:00 A.M.: BISMARCK salvos: BISMARCK position sometimes in lee of PRINZ EUGEN (see various photos, film):
1. HOOD ahead
2. Between HOOD and PRINCE OF WALES
3. covering HOOD (fire observed: boat deck)—questionable: of this salvo, at most one 38.1-cm shell may have hit HOOD—presumed 20.3-cm hit from PRINZ EUGEN—3rd salvo: according to NORFOLK (Phillips) observation: Hit on upper deck, torpedo tubes . . . Explosion, gunpowder (cordite) fire (red-orange, black and yellow smoke)-according to Tilburn (HOOD) observation: port side hit amidships (near rocket bodies against low-flyer very near 10-ton 10.2-cm ready ammunition—Dundas (HOOD) observation: Cordite fire on boat deck (starboard side)—Report to bridge (Briggs/HOOD confirmation)—Vice-Admiral HOOD (on fire in ready ammunition: "Leave it until the ammunition is gone!"
4: Aft from HOOD
5: See 6:01 A.M.
2nd PRINZ EUGEN salvo (see 5:55 A.M.: 1st PRINZ EUGEN salvo): "Fire four!—four-hectometer straddling group!:, 2nd salvo (situation): 4 hits ' Fire standing salvo! Hit at level of second funnel and after mast (quickly spreading fire). Both units (BISMARCK and PRINZ EUGEN) are on target after first salvos—after the 5th and 6th PRINZ EUGEN salvos: Change target to PRINCE OF WALES (Signal from Fleet Chief, BIS-MARCK)—Two hits are observed there.

ca. 5:56-5:57 A.M.: PRINZ EUGEN scores a hit on HOOD with the 4th salvo (height of hangar, after mast): Fire near ammunition lift (port, aft) . . . fire spreads . . .
ca. 5:58-5:59 A.M.: HOOD fire quickly dies down-Lütjens command (after 2nd or 3rd BISMARCK salvo, after 5th (6th) PRINZ EUGEN salvo): Change target to left! (against PRINCE OF WALES).
ca. 5:59-6:00 A.M.: BISMARCK-PRINZ EUGEN: "Change target to left!" (no target correction needed: distance/situation-enemy's course remains almost the same).
5:59 A.M.: PRINZ EUGEN: Beginning of effective firing (after getting range)—Observed effect: two hits, small fire-distance about 160 hm—use of heavy (10.5-cm) anti-aircraft guns (with decreasing distance).
ca. 6:00 A.M.: PRINZ EUGEN observation: Enemy salvo in wake . . .
ca. 6:00 A.M.: Position of Force H between Cape Trafalgar and Cape St. Vincent-Home Fleet position south of Iceland (distance from site of HOOD sinking position: approximately 550 nautical miles).
6:00 A.M.: HOOD changes course (?) by 20 degrees (to port?), to bring the after turrets (C, D) into action—Speed: about 28 knots—distance from enemy: about 197 hm (MDV No. 601).
6:01 A.M.: Distance to HOOD about 135-140 hm (about 16,500 yards ??) 5th BISMARCK full salvo (the 3 (I to III) port 15.0-cm

double turrets fire on PRINCE OF WALES) against the HOOD, approaching (?) to port at high speed—the decisive (penetrating side armor) hit (BISMARCK armor-breaking shell, or two BISMARCK armor-breaking shells—ignition setting not known, muzzle velocity: 829-957.79 meters per second) on the boat deck near the mainmast (location almost identical to effect of 3rd BISMARCK salvo)—On HOOD the 5th salvo (Turret C or D?) had just been fired (?)—two hits (by heavy artillery shells: water-column height about 30-35 to 40 meters ' measures taken to correct targeting) can also be observed very near HOOD-flames between the HOOD's masts, massive flames at height of Turrets C, D (about 200 to 300 meters) together (struck: after 10.2-cm ammunition chamber . . . about 20 meters farther . . . near mainmast abaft: presumed explosion of the 38.1-cm cartridge chamber, contents some 112 tons of gunpowder) with a white-glowing (briefly circular) ball/cone/mushroom (lasting 1/3 of a second!)—Heavy cloud of smoke on HOOD that flies into the air! (Bow and stern now separate parts?): HOOD is torn apart (high columns of metal parts)—bow briefly visible pointing out of the water (about 30 degrees) . . . central portion of HOOD breaks up--heavy artillery Turret C or D (?) falls into water at great distance to left of ship, continued smoking on the side hit—sinking position: approximately 63 degrees 20 minutes north, 31 degrees 50 minutes west (maritime chart: AD-73)—PRINCE OF WALES moves close (with opposed rudder) on south course (danger of collision with HOOD ruins and debris—to a short distance from HOOD, thus only slight German target correction necessary, 1 to 2 degrees) . . . Further observations of HOOD: from PRINCE OF WALES

(Terry/Paton: as HOOD rolls over to port, starboard ribs and jagged stern portion are seen—also (immediately after explosion) large pieces of ship in the air: mainmast, 38.1-cm gun barrel, main crane, etc.: "A heavy black pall of smoke covers the ship as it sinks by the stern and turns 180 degrees in the process."—HOOD crew: 95 officers and 1324 men—there were only three HOOD survivors, who could be rescued after three hours by the British destroyer ELECTRA:

1. Ordinary Seaman (Signal Corporal) A. Edward Briggs . . . later Lieutenant. Station: Compass deck, navigation bridge—Command messenger of Flag Lieutenant of Vice-Admiral Holland.
2. Midshipman (Ensign) William J. Dundas . . . later Lieutenant Commander. Station: Upper (enclosed) bridge, as Ensign of Watch, under command of Officer of Watch.
3. Able Seaman (Senior Corporal) Robert E. Tilburn . . . later Admiralty Storeman. Station: Boat deck, service, port 10.2-cm anti-aircraft guns.

After 6:01 A.M.: The PRINCE OF WALES, moving in close starboard formation to HOOD, turns hard around the HOOD wreckage and rubble and moves off on southward course through the fog and black smoke of concentrated fire from BISMARCK and PRINZ EUGEN.

As of 6:01 A.M.: BISMARCK and PRINCE OF WALES change target--distance about 148 hm.

6:02 A.M.: PRINCE OF WALES under concentrated fire of BISMARCK and PRINZ EUGEN, shooting well and fast (distance about 160 hm)-NORFOLK fires 3 salvos (SUFFOLK six in all?).

Between 6:03 and 6:14 A.M.: PRINZ EUGEN—acoustic surveillance (listening area) : 3 torpedo courses (unexplained: sound at 270 degrees, among others, torpedo course at 220 degrees, among others, distance about 165 hm): "Torpedo sounds to

port!" (No explanation of these torpedo courses possible, possibly aircraft?): Rudder hard-a-starboard (signal to BISMARCK).

6:03 A.M.: PRINCE OF WALES takes direct hit in the bridge and leaves the battlefield smoking heavily.

As of 6:03 A.M.: Rear Admiral Walker takes over further measures as Senior Commander (AK).

Approx. 6:05 A.M.: PRINCE OF WALES fire near BISMARCK (also possible long-range fire at PRINZ EUGEN: one 15-cm splinter of a 35.6-cm shell was found on the after port side deck) as well as (from aft . . . distance: about 4000-5000 meters) 20 3-cm shots from NORFOLK . . . Air alarm ("Catalina" flying boat . . . possible torpedo planes, torpedo courses??)—Two hits on PRINZ EUGEN are observed on PRINCE OF WALES (shot observation difficult)—distance about 140 hm (PRINCE OF WALES surrounded by 38.1-cm, 20.3-cm and 10.5-cm shots (shot sequence: every 10 to 15 seconds), departs smoking and burning: PRINCE OF WALES was hit by seven shells): 4 (BISMARCK) 38.0 cm, 3 (PRINZ EUGEN) 20.3 cm: 38.0-cm hit the bridge, 38.0 cm superstructures, forward medium artillery control post, 38.0 cm under the waterline near diesel generator room, 20.3 cm at waterline aft, ca. 500 tons of water, 20.3 cm at waterline aft in the ship, 20.3 cm in 13.2-cm shot loading room.

6:09 A.M.: The last salvos: German fire ceases about 195-215-220 hm—When the battle began—for sufficient tactical reasons—the lightly armored PRINZ EUGEN remained in "the line" (was not ordered into the lee of fire), presumably because the British units, quickly arriving at a sharp angle, at first were said to be cruisers or destroyers. On this subject, Group Command North: ". . . one cannot think schematically and must take

danger in the bargain. Dangers-as here—do not necessarily mean losses or defeat."—BISMARCK took three hits (among other results, reduction of speed) . . . Plan: return to St. Nazaire or Brest): 2 heavy and one light hits.

1. Sections XIII-XIV ' electric system out of commission—4-boiler room port 2 makes water (can be held): possible oil loss?
2. Forward sections XX-XXI: entered at port, exited starboard over armor deck—oil cells hit: light port side damage (about 4000 tons of water in ship).
3. Hit through a boat (without importance).

Effects:
a) Reduction of speed (top speed cut to 29 knots . . . water in ship forward
b) wide, clearly visible oil spill.

Ammunition consumption:
BISMARCK: 93 38.0 cm
PRINZ EUGEN: 183 20.3 cm
(Comparatively high, especially for PRINZ EUGEN).

6:10-9:00 A.M.: Weather: Northeast wind, strength 2 to 3, sea 3, visibility clear—sunny—15 nautical miles, swells from south. Fogbanks as of 11:00 A.M.

Battleship BISMARCK

Battle Cruiser HOOD

Heavy Cruiser PRINZ EUGEN

Right: the last picture of the British battle cruiser HOOD, taken from the PRINCE OF WALES while running in formation to starboard of flagship. In the foreground the guns of the 35.6-cm quadruple turret of the PRINCE OF WALES.

Next page: Battle Cruiser HOOD lying at anchor in Scapa Flow, 1941 . . .

Saturday, May 24, 1941, between 5:35 and 5:55 A.M. (board time: 4:35 to 4:55 A.M.): a few minutes before the battle began: BISMARCK seen from PRINZ EUGEN . . .

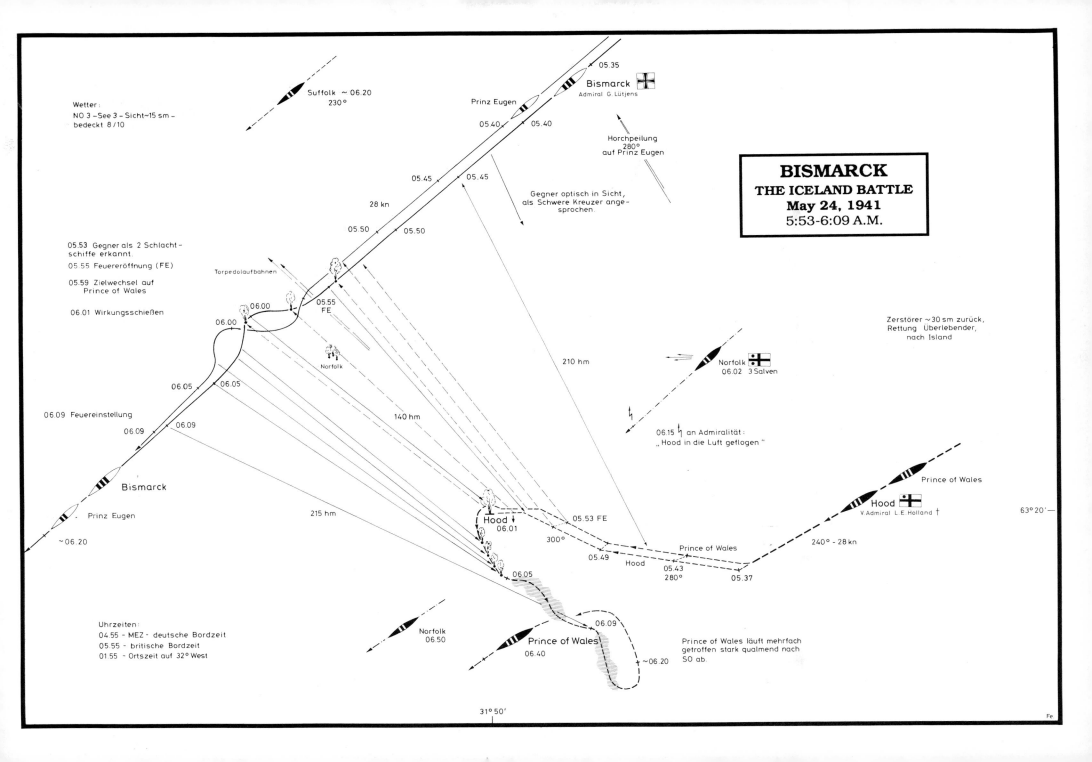

BISMARCK
THE ICELAND BATTLE
May 24, 1941
5:53-6:09 A.M.

Wetter:
NO 3 –See 3 –Sicht~15 sm –
bedeckt 8/10

Suffolk ~06.20
230°

Prinz Eugen

05.35
Bismarck
Admiral G. Lütjens

05.40   05.40

Horchpeilung
280°
auf Prinz Eugen

05.45   05.45

28 kn

Gegner optisch in Sicht,
als Schwere Kreuzer ange-
sprochen.

05.50   05.50

05.53 Gegner als 2 Schlachtschiffe erkannt.

05.55 Feuereröffnung (FE)

05.59 Zielwechsel auf
Prince of Wales

Torpedolaufbahnen

06.01 Wirkungsschießen

06.00   05.55
FE

06.00

06.00

210 hm

Zerstörer ~30 sm zurück,
Rettung Überlebender,
nach Island

Norfolk

Norfolk
06.02 3 Salven

06.05   06.05

06.09 Feuereinstellung

140 hm

06.15 ⚡ an Admiralität:
„Hood in die Luft geflogen"

06.09   06.09

215 hm

Bismarck

Prinz Eugen

~06.20

Prince of Wales

Hood
06.01

05.53 FE

300°

05.49

Hood

05.43
280°

05.37

Prince of Wales

240° - 28 kn

Hood
V. Admiral L. E. Holland †

63°20'—

Uhrzeiten:
04.55 – MEZ – deutsche Bordzeit
05.55 – britische Bordzeit
01.55 – Ortszeit auf 32° West

Norfolk
06.50

06.09

06.05

Prince of Wales
06.40

~06.20

Prince of Wales läuft mehrfach
getroffen stark qualmend nach
SO ab.

31°50'

Fe

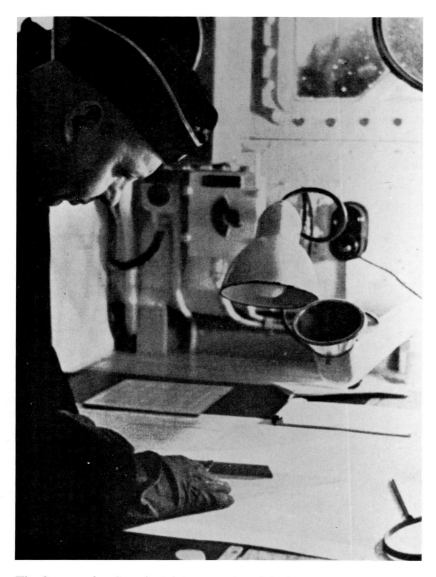

The Commander, Captain Brinkmann, aboard the heavy cruiser PRINZ EUGEN, at the map table in the command post. At right over the table is the speaking tube to the bridge, at upper right a rotating panel that cleans the glass for good vision in rain or snow.

114

Saturday, May 24, 1941, between 5:35 and 5:55 A.M. (board time: 4:35 to 4:55 A.M.)—on the bridge of the heavy cruiser PRINZ EUGEN, with binoculars, is the Commander, Captain Brinkmann: "The Commander observes the approach of the British battle group."

The heavy turning hoods with the range-finding beams and electronic antennas on the forward and after command posts as well as on the fighting top are the most noticeable indications of the sea-target fire control facilities on the BISMARCK.

The range-finder beams had lengths of ten and seven meters (the latter on the front command post). The rectangular frame antenna measured about 3.5 by 1.5 meters.

125

According to the conditions of the time, the BISMARCK was optimally equipped with optical and mechanical fire-control apparatus. Even her former enemies maintain that the fire-control equipment, particularly the optical devices, on German ships were qualitatively ahead of those on the warships of other nations. This is true to the extent that the German devices made better use of residual light. But this capability was limited by such factors as fog and misty backgrounds. It is shown from the events of the war, completely independent of the question of whether German fire control was better than that of their enemies, that the Allied warships scored equally good firing results. During the exchange of salvos between the BISMARCK and the HOOD, the British controlled their fire just as accurately as the Germans. If the BISMARCK had escaped its early death, it surely would have lost its *de facto* superiority to its opponents' radar equipment.

The results of the BISMARCK operation show that, unlike the German Navy, the British recognized and fully utilized the possibilities of this new means of control. While the British reconnaissance units were able to shadow the BISMARCK at radar range and thus knew her location, the German fleet command was usually in the dark as to the whereabouts of their opponents' scouts. It would be incorrect to attribute this lack to the sensitivity of the German devices to disturbances, since the same weakness existed in the other side's radar equipment.

As of 1941 the German battleships had lost their freedom of movement because of the effective use of British electronic surveillance equipment (radar).

What remains of the technical fascination of the BISMARCK must take a secondary place to the avoidable death of more than 2000 German naval men.

Above: the after range finder from another angle. The casing for the instrument had side armor 100 mm thick; the cover measured 50 and the bottom 30 mm.

Three optical systems for fire control can be seen in this picture of Turret "Berta." On the diagonal part of the cover, a periscope is visible; one was also installed on the opposite side. They help the gunners to watch the battle. Here it is covered with a protector. In addition, for every gun there was an aiming telescope available. This means of fire control was used only for short-range fighting; to use it, the firing-post shutters, made of armor steel, had to be pushed away from the firing openings. Otherwise the openings remained closed when at sea to keep sea water from getting inside the turrets. The housing that projects from the turret end covers the ten-meter basic device that is built into the turret. This optical device, like all other optical apparatus, is supposed to be stabilized. Of course it can only be partly stabilized, since it is firmly attached to the turret and can only equalize, with the turret, lateral and vertical movements. The rolling axis is not stabilized.

Left page, below:
This is the electric range-finder position on the fighting top, likewise equipped with a range-finder antenna. In order to reduce topheaviness, the armor thicknesses were kept low, with just 20 mm for the cover, 60 mm for the side walls and 20 mm for the bottom.

These range finders rank among the most capable optical devices of their time. They worked according to the mixed-picture process. Two identical pictures of the observed object appear to the observer in the eyepiece. The correct distance to the object in question can be read on a scale when the two images are positioned over each other by turning a hand wheel. The longer the measuring beams were, the more exact the results of measuring were and the greater the measured distances. Through loss of light in the lens system, tolerance limits, friction losses in the mechanical elements and other factors, technical limits were set for further improvement of the performance of these devices. In the end, though, weather and light conditions set the limits of their usefulness.

Just as there were for sea-target fighting, there were also special range finders for use against air targets. The picture shows the port anti-aircraft range finder under a protective cover (ball anti-aircraft position). Since the measuring distances were considerably shorter than those for sea targets, the measuring beams of the anti-aircraft range finders could be made shorter than those for use on sea targets, and were four meters long. There was also a tactical reason for this, for the higher speeds of the air targets required a more moveable and therefore simpler process. The range finder in the third dimension was given high priority. Very differently from sea-target range finding, in that of air targets a large measuring error occurs when the angle cannot be equalized. This measuring error is carried over to the anti-aircraft weapons as an aiming error which they receive from their aiming values.

To optimalize aiming, the basic device was thus cardanically mounted. Since it was firmly attached to the protective hemisphere, this too had to be horizontally adjusted.

The striking hemispherical construction thus remains independent, in terms of aiming stability, of how the ship moves under it.

Only the two forward anti-aircraft range finders are protected by hemispherical covers—the pictures on the right page show the two unprotected range-finding positions, at left toward the stern and at right on the boat deck, which were often referred to as auxiliary positions. This term is incorrect because they were meant to be just as valid as the apparatus on the turrets in serving fire control against air targets toward the stern. When several air targets approached the ship's sides simultaneously, fire control had to be divided between the range finders forward and aft. All the same, their construction does not correspond to the standards of those used on the TIRPITZ. What happened here is that the full-value anti-aircraft range-finding equipment was done without to follow the treaty agreements with the Soviets, who had obtained a heavy cruiser of the German construction type that had been equipped with spherical anti-aircraft fire control posts that were actually reserved for the BISMARCK.

The open anti-aircraft range-finding position toward the stern. The optical devices have the same construction features as the ten-meter measuring beams and also work by the mixed-picture process.

Below: This four-meter measuring beam is part of the open position on the after boat deck. In the foreground are speaking tubes—one of the means of transmitting commands.

It is understandable that environmental conditions had to have a negative effect on these fire control devices. The psychological influence of a missed shot on the personnel operating them also must not be underestimated.

Right:
One of the classic means of lighting sea space is still the floodlight. This fire control aid is still valued too highly in the navy. German ship commanders and artillery observers still rely on "white" light, though British ships are already navigated with the help of electronic waves. The overvaluing of the floodlight by the Germans results in expensive apparatus. Because of the inherent weight involved, these devices could no longer be turned manually, but had to be equipped with remote controls. The picture on the next page makes the laborious design clear. The individual components such as drives, floodlight covers, power sources, etc., are particularly easy to see. This floodlight is mounted in the crow's nest of the fighting top. The equipment was photographed from inside the top through the bulkhead opening. The heavy armor plates are also noteworthy.

This picture shows the means of stabilizing. The floodlight moves cardanically between the mounting arms (pitching), on the socket arm (rolling) and the socket itself (yawing).

This is the floodlight platform beside the funnel, with its two floodlights, as seen from aft. The navy was urged throughout the war to use this equipment to aid in fire control, even though their presence had to be reduced in favor of light anti-aircraft weapons.

The Allies, on the other hand, would have fire-control radar available in little more than a year.

At each side of the funnel there was also a floodlight set up in a crow's nest. They were protected from shrapnel and weather damage by round covers. The warmth of the funnel, isolated by the closed covers, also kept the moving parts of the devices ice-free. Thus their readiness for use was fully guaranteed in any weather. These stations could be reached quickly over a bridge from the fighting top.

The turning hoods with the basic devices had not yet been built in.

The crow's nests on the fighting top, on the funnel and the boat deck are all equipped with floodlights already. The forward anti-aircraft range-finding post under the hemispherical protector has also been installed.

This picture and those on the next two pages show equipment that probably was only photographed once on the BISMARCK. Because of the secrecy that applied to these rooms of the ship, they could be photographed only with official permission.

These are command and calculation headquarters. The picture at left shows loudspeakers on the walls for transmitting commands and, in the right corner, a calculator. It could be possible that the ladder leads through an armored shaft to a 15-cm turret.

In the next picture there is a radar navigation or evaluation room. Over the device hang the cables for means of communication, and at the right are calculator units for a sea-target reckoning position. Further identification could lead to false statements, for most of the men who built and used this equipment are no longer alive. Secrecy regulations also prevented anyone being informed beyond the realm of his own work.

Despite this, one can put oneself in the place of these servicemen:

Here they work behind steel walls, thoroughly screened from outside influences. The sounds of the machinery are noticeable here only as a soft humming and vibration. Here the sterility, concentration, precision and responsibility of laboratories prevail, even intensified by the speed with which vital decisions must be made. Here a mechanism is used that was only replaced by electronic processes slowly, beginning in the early Fifties.

135

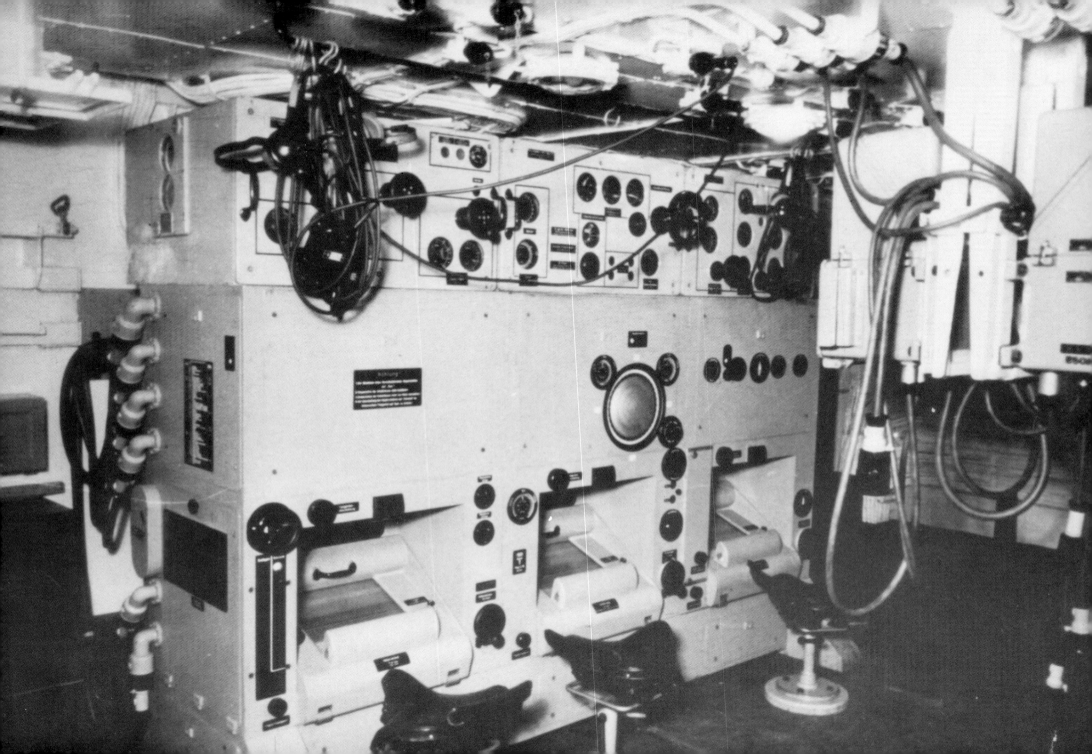

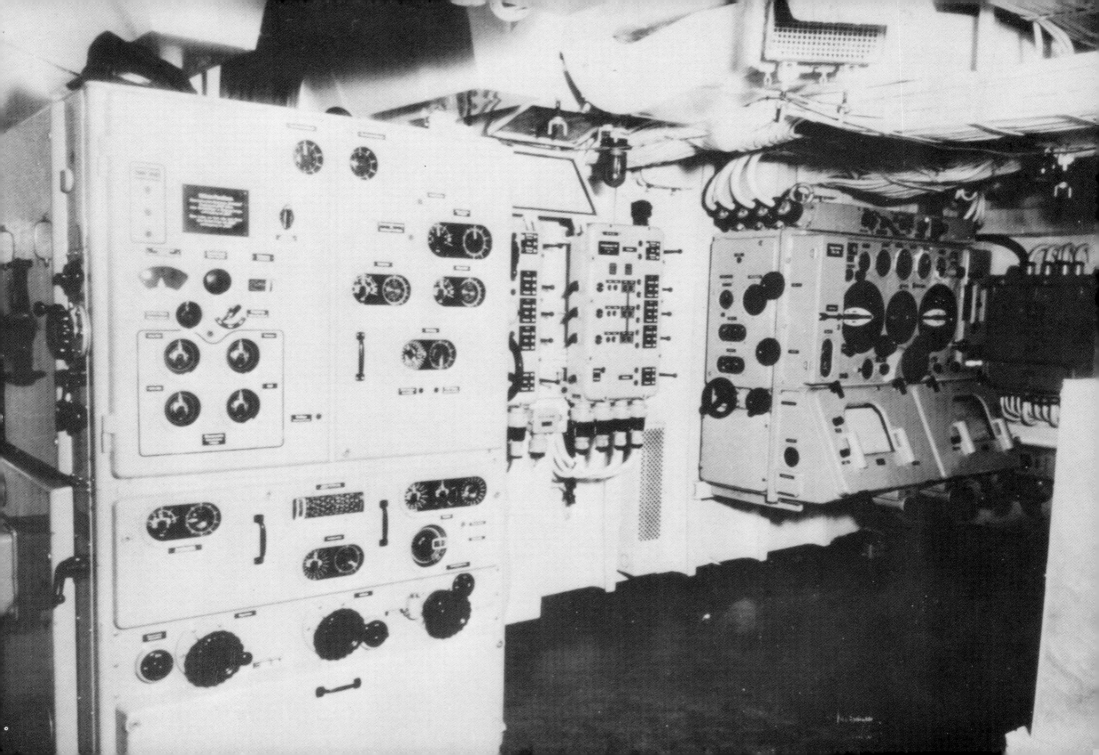

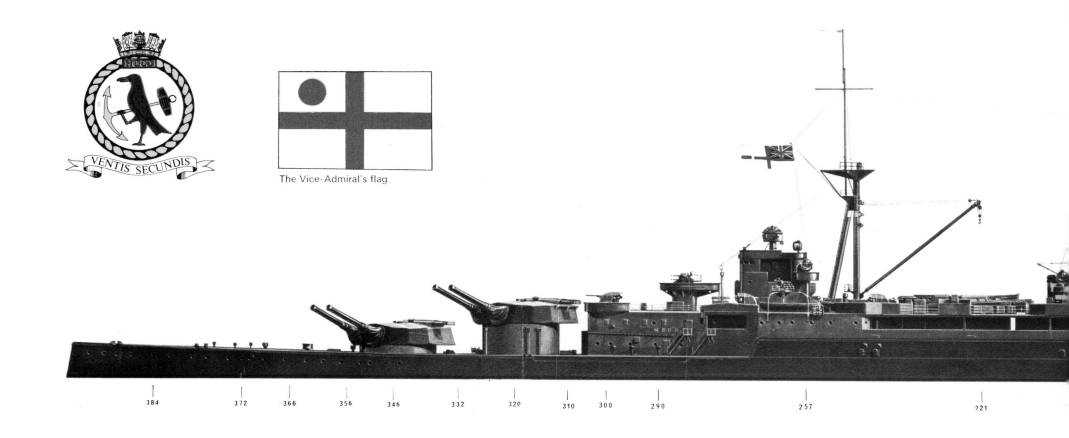

VENTIS SECUNDIS

384   372   366   356   346   332   320   310   300   290   257   221

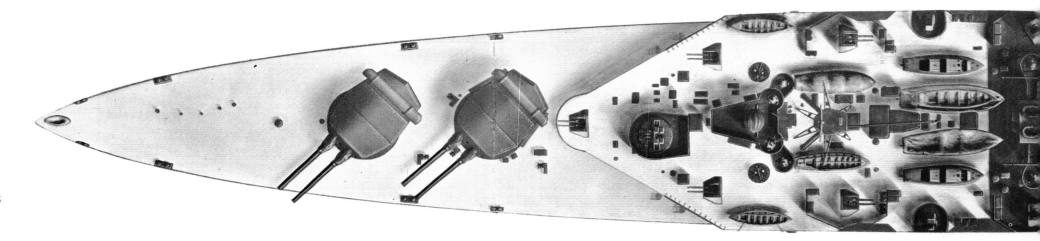

138

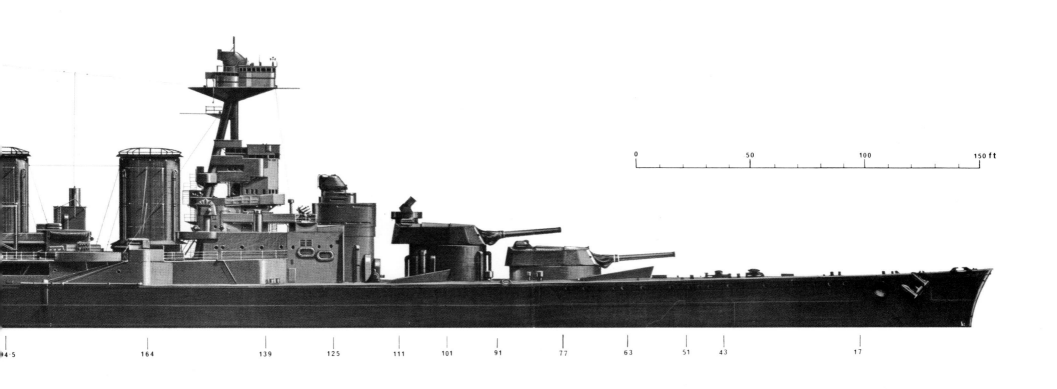

0   50   100   150 ft

184·5   164   139   125   111   101   91   77   63   51   43   17

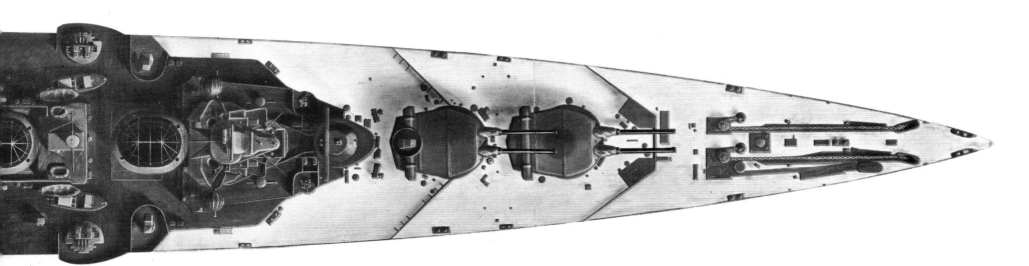

139

British Torpedo Planes Attack the BISMARCK

On May 24, about 7:00 P.M., about thirteen hours after the battle with the HOOD group, the PRINZ EUGEN was able to separate from the German battleship, at first unnoticed. The BISMARCK was now alone and knowingly drew all the naval powers of the enemy toward itself. Its heavy and fully battle-ready units had not yet closed in to within fighting range of the BISMARCK, though. Admiral Tovey decided to engage his torpedo planes, which had the range to reach the BISMARCK and could try to reduce its speed by scoring hits on it. The aircraft carrier H.M.S. VICTORIOUS was available. This was a modern carrier that had been put in service only in 1941. With a displacement of 23,000 tons, it could carry 72 planes. Two catapults and two lifts kept the planes flying. Its armament consisted of 16 11.4-cm anti-aircraft guns in twin mounts and 64 4- and 2-cm anti-aircraft guns (as of 1944). With 11,000-HP performance, it could attain 31 knots. Its crew numbered 1600 men.

The picture on the left page is one of the great historic photo documents of World War II. It shows the nine "Swordfish" torpedo bombers of the 825th Squadron, under the command of Lieutenant Commander (A) Eugene Esmonde, on the rainy flight deck of the VICTORIOUS. The wings, of course, are not yet folded, but it was almost time for the first attack. One of the braking ropes is visible running across the flight deck. It was raised during landings so that the hook on the fuselage of a plane could catch on it.

Here is a "Swordfish" dropping an 18-inch torpedo. This model was one of the most successful sea-fighting planes. This capability probably originated with the aircrews of the Royal Navy who then flew the already obsolete "Swordfish" into battle with coolness and determination. The torpedoes had to be flown toward the ship target by the carrier plane before being dropped. If the attack was to be successful, the torpedo had to be dropped only a few hundred meters from the target.

Technical data (1934 form):

Fairly Swordfish—ship-supported multipurpose aircraft, powerplant: Bristol-Pegasus radial engine producing 750 HP—Weights: dry weight 2410 kp, takeoff weight 3685 kp, top speed 220 kph.
Armament:
One fixed 7.7-mm machine gun on the side of the upper front fuselage; one movable 7.7-mm machine gun in an open mount on top of the fuselage—Load to be dropped: about 750 kp torpedo or bombs; crew: two men.

Between 1935 and 1945 some 2400 of these planes were built. Throughout the war this model was used as a torpedo, bomber, reconnaissance, smoke-screen and illumination plane.

The MODOC, a Coast Guard patrol boat of the TAMPA class, was an unwelcome "guest" in the first aircraft attack on the BISMARCK. The 1780-ton craft (built 1921, 16 knots, with two 12.7-cm and two 7.6-cm guns) moved into the area of the attacking torpedo planes while on a Greenland patrol.
Ships of this size were mentioned only rarely in war reports. Generally, such classes of ships did not even have official names, but were differentiated only by code numbers. All the same, they were to gain in significance increasingly during the further course of World War II.

Between 10:30 P.M. and midnight the "Swordfishes" attacked the German battleship in three flights of three planes each and several "Fulmars" (six machines aboard the VICTORIOUS). There are no documentary photographs of this attack.

Thus did Walter Zeeden (1891-1961) envision the attack of two "Swordfishes" in a 1941 painting.

Between 5:55 and 6:09 A.M.: Starboard side, upper deck (aft) of **PRINZ EUGEN** (starboard 3rd 10.5-cm double anti-aircraft guns):
"Two shots from the 38-cm double turrets of the **HOOD** land close to **PRINZ EUGEN**."

Between 5:55 and 6:09 A.M.: View from **PRINZ EUGEN** (starboard side, upper deck) past the C and D (aft) turrets sticking out to port, in the direction of **BISMARCK** (almost in line with the keel, thus first phase of battle): :Battleship **BISMARCK** fires the first salvo on the battle cruiser **HOOD**."

Between 5:55 and 6:09 A.M.: BISMARCK swings to starboard (still firing from port aft). Before the new plume of smoke blowing away (right) a 35.6-cm shell from PRINCE OF WALES has fallen.

Between 5:55 and 6:09 A.M.: BISMARCK swings to starboard—left: a 35.6-cm shell from PRINCE OF WALES has fallen.

Between 5:55 and 6:09 A.M.: BISMARCK turns (again) to port: "The fast salvo rate hides it at times in thick powder smoke."

Between 5:55 and 6:09 A.M.: The BISMARCK turns to port (again, firing from port side).

A few seconds before 6:01: (Board time: 5:01 A.M.): the British battle cruiser HOOD, shaken by a
38.0 full salvo from the BISMARCK (distance about 155 hm), shortly before the decisive explosion
(maximum time 1/3 second!). Time between 06:00:57 / 06:00:58 / 06:00:59 / 06:01:00 / 06:01:01 /
06:01:02 A.M.

Able Seaman Robert E. Tilburn, one of the three HOOD crewmen rescued—1938: as Boy 2nd Class entering the Royal Navy— 5/24/1941: HOOD station: Boat deck (port side) 10.2-cm anti-aircraft guns. Saw (presumably PRINZ EUGEN's) port hit (amidships). When the HOOD sank he was at first pulled down into the water and rescued after about four hours by the destroyer ELECTRA (Commander Cecil Wakeford May).
—After World War II: Admiralty Storeman.

After 6:01 A.M. (board time 5:01): Sinking place of HOOD (approximately 63 degrees 20 minutes north, 31 degrees 50 minutes west), with smoking HOOD wreckage (presumably after 38.)-cm double turret), far left at edge of picture: :The German ships are now firing on the retreating PRINCE OF WALES, behind whose stern the shots can be seen to have fallen.''

About 6:09 A.M. (board time: about 5:09)—the last heavy artillery double salvo (Turrets C & D) fired at the PRINCE OF WALES by the BISMARCK, at about 215 hm (photographed from PRINZ EUGEN in the lee of fire)—PRINCE OF WALES (which ceased fire at 6:21) received a total of seven hits (four 38.1 from BISMARCK, three 20.3 from PRINZ EUGEN)—Ammunition used by the Germans against the HOOD and PRINCE OF WALES: BISMARCK 93 38.0, PRINZ EUGEN 183 20.3—

BISMARCK was hit three times: 1 in the forward bow—Section XXI (Sections XX and XXI filled with water)—2 in Sections XIII-XIV—3. hit the upper deck (results: decreased speed, clearly visible oil leak). Shortly before this picture was taken, BISMARCK turned out of PRINZ EUGEN's wake and passed close to the heavy cruiser, which was turning slightly to starboard. The BISMARCK's heavy artillery turrets remained in this (last) firing position.

Just afterward—about 6:11 to 6:15 A.M. (board time 5:11 to 5:15), two more pictures were taken from the PRINZ EUGEN (the BISMARCK's heavy artillery was still in the same position—compare the photo of the last salvo, about 6:09 A.M.!)—6:15 A.M.: NORFOLK reports: "Hood blew up! . . ."
"Number change: BISMARCK goes ahead of PRINZ EUGEN." In the process, BISMARCK turned into the PRINZ EUGEN's lee, then somewhat later passed the heavy cruiser (compare the two following photos).

6:32 A.M.: BISMARCK reports to Group North: "Sank battle cruiser, presumably HOOD. Another battleship, KING GEORGE or RENOWN, turned back damaged. Two heavy cruisers stay in touch . . ." Signal from Fleet Chief to PRINZ EUGEN (another request to "change numbers"), whereby the flagship approaches PRINZ EUGEN at high speed. "In the maintop is the admiral's flag of the Fleet Chief."

The BISMARCK passes the PRINZ EUGEN (the heavy artillery turrets A & B, C & D are again in "zero position").

PRINZ EUGEN's Commander, Captain Brinkmann, examines a piece of a 38.1-cm shell from the HOOD (long-range shot) that was discovered by chance on the after port deck by LI (Frigate Captain (Eng.) Graser).

Between 3:00 and 6:14-6:34 **P.M.**: **BISMARCK** in **PRINZ EUGEN**'s wake. The last photo of the **BISMARCK** in the evening after the battle.

Fire Control

The modern battleships of the last generation attained weapons performance with which sea targets could be attacked even at a distance of 400 hm. In the process, speeds in departing or approaching courses up to 100 kph, this being almost 30 meters per second, had to be taken into consideration. There was also a danger from the air, effective from heights of some 4000 meters and flying speeds around 150 meters per second.

This performance required tactical decisions within such short times that the ships' commanders were provided with technical assistance to improve the likelihood of hitting the target. The effect and the value of all naval weapons was, in the end, determined by their ability to hit targets.

Provided that the artillery equipment is in faultless condition, the target-hitting performance depends on the exactness of one value. That is the "allowance" to the side, and in fighting air targets also the "allowance" in terms of height. The "allowance" is the positioning of the gun barrel far enough ahead of and/or above the moving target that the fired shot hits the intended target because a number of single factors have been taken into consideration.

Decisive for determining this allowance are, among others, values such as the

—Speed and direction of the target
—Speed and course of the ship itself
—Distance to the target
—Wear of the gun barrel
—Type of ammunition
—Muzzle velocity
—Powder temperature
—Weather influences
—Correction of the angle of descent.

The fire control equipment could already deal with these values so precisely then that the possible firing distances could be utilized, so that for firing under favorable conditions, corrections of only a few hundred meters were attempted.

German industry had then attained a level that enabled it to equip the German ships with the most modern mechanical and optical fire-control equipment. Their characteristics essentially determined the external appearance of the heavy German naval units.

The highest optical achievement of the BISMARCK was attained by the 10-meter range finder (Basic Device BG) of the artillery command post, which was attached to the foretop 27 meters above the surface of the water. From here the visual distances in all directions could be utilized optimally without limitation by powder smoke or funnel exhaust. Disturbances in movement by the ship's hull were neutralized by the component-stabilized device.

The artillery command post was used like panoramic radar; with it the First Artillery Officer apportioned and directed the action of all sea-target weapons. The conduct of the battle against sea targets could then be taken over from two command posts. The main post, with a 7-meter basic device, was on the bridge, and the second post, with a 10-meter device, was toward the stern. Both posts had access to their appropriate fire-control headquarters through armored shafts. Here all information for fire control was collected and coordinated. Fire-control calculators were available for assistance. The calculating speed of these "computers" was remarkably slow by today's standards. Every value had to be put into these shot-value calculators manually, and mechanical equipment processed the entered values. There were devices for various kinds of basic calculation, plus angle-function devices, differentiation and integration devices to determine minimal and maximal values, and coordinate changers. Unchanging figures, such as data from the shot tables, were "stored" in the form of curved bodies. According to the ship's fighting situation and condition (damage), control could be switched from one command post to the other, and the turrets for the heavy or medium artillery could be directed independently and supplied with information from various posts.

The fire control system on the BISMARCK was redundant to the extent that every heavy artillery turret was equipped with a ten-meter basic device, and could take over sea-target fighting individually if the central fire control was put out of action. One exception was the forward turret, from which the apparatus had unfortunately been removed on account of damage from sea water.

Like the apparatus in the foretop, all basic devices had stabilized components. Even the central gun turrets of the medium artillery were equipped with component-stabilized 7-meter basic devices.

The anti-aircraft guns were directed by two forward and two aft 4-meter anti-aircraft control posts. The two forward posts were protected by shrapnel-proof hoods (ball AA control posts), in which the measuring devices were mounted cardanically. Also part of the anti-aircraft fire control system were stabilized target-indicating devices for apportioning air targets among the individual AA batteries.

At that time, floodlights were still components of the fire control system for air and nearby sea targets. The BISMARCK had: one floodlight on the crow's nest of the fighting top, two on the light platform of the funnel, two in crow's nests on the front of the funnel, and lastly two on the after boat deck behind the mainmast. The floodlights were also stabilized on three axes and could be remote-controlled and turned on targets by floodlight aiming columns.

During night fighting at greater distances, the target area was lit by flares to provide the necessary light for the optical fire control facilities.

Some publications suggest that the BISMARCK was already equipped with a radar surveillance device (fire control radar). This is incorrect; the characteristic construction of the antenna grid masts, as they were attached to the turning hoods of the foretop and the forward and after command posts, are only an indication of the ship's being heavily equipped with devices for seaspace surveillance and range finding, so-called electronic navigation instruments (FuMos).

Only the pressure of events led to these devices being used in a limited way for fire control on other vessels later.

In the attack, 18 torpedoes were dropped against the BISMARCK. According to this, since the "Fulmars" were righter planes, the "Swordfishes" must have attacked with two torpedoes each. That was an extraordinary procedure, since these machines were normally equipped with only the hanging gear for one torpedo.

The "Swordfish" of Lieutenant Gick and Air Gunner Petty Officer (A) L. D. Sayer scored with the eighteenth and last torpedo dropped, at about 11:38 P.M., in rainy weather and heavy wind of force 7 to 8, hitting the belt armor in the area of the foremast on the starboard side to score the only hit. The air photo on the previous page probably shows the clouds of smoke from this torpedo hit. The effect of the approximately 300 kp of high explosive contained in the torpedo certainly could endanger a warship the size of a battleship. But for the BISMARCK this blow was of no significance.

Returning "Swordfishes" prepare to land. This picture was not taken in connection with the May 24-25 attacks. In these operations only two "Swordfishes" and one "Fulmar" plane were lost. The crew of the "Fulmar" could be rescued a few hours later, one "Swordfish" crew only after nine (!) days; the other torpedo-plane crew remained lost.

It is noteworthy that a large aircraft carrier like the VICTORIOUS with a capacity of 72 planes obviously had only 15 battle-ready planes and their crews.

This is an indication of how ill-equipped in personnel and material the British were for this type of sea warfare at that time.

This photo shows some of the crewmen of the torpedo planes, gathered here aboard the **ARK ROYAL** to be decorated for their action against the **BISMARCK**. In action against individual targets such as ships, the coolness, courage and determination of individual men were often decisive.

From left to right: Lieutenant Percy D. Gick, D.S.C.; Lieutenant Commander (A) Eugene Esmonde, D.S.O.; Sub-Lieutenant V. K. Norfolk, D.S.C.; Air Gunner Petty Officer (A) L. D. Sayer, D.S.M. (he flew in Gick's plane that registered the hit), and Leading Airman A. L. Johnson. Eugene Esmonde was lost on February 12, 1942 in an attack on the **SCHARNHORST, GNEISENAU** and **PRINZ EUGEN** during the Channel breakthrough.

The German naval men on the **BISMARCK** were surprised at the high morale of the attacking British. The German National Socialist war propaganda had tried for a long time to portray the British soldier as soft and decadent.

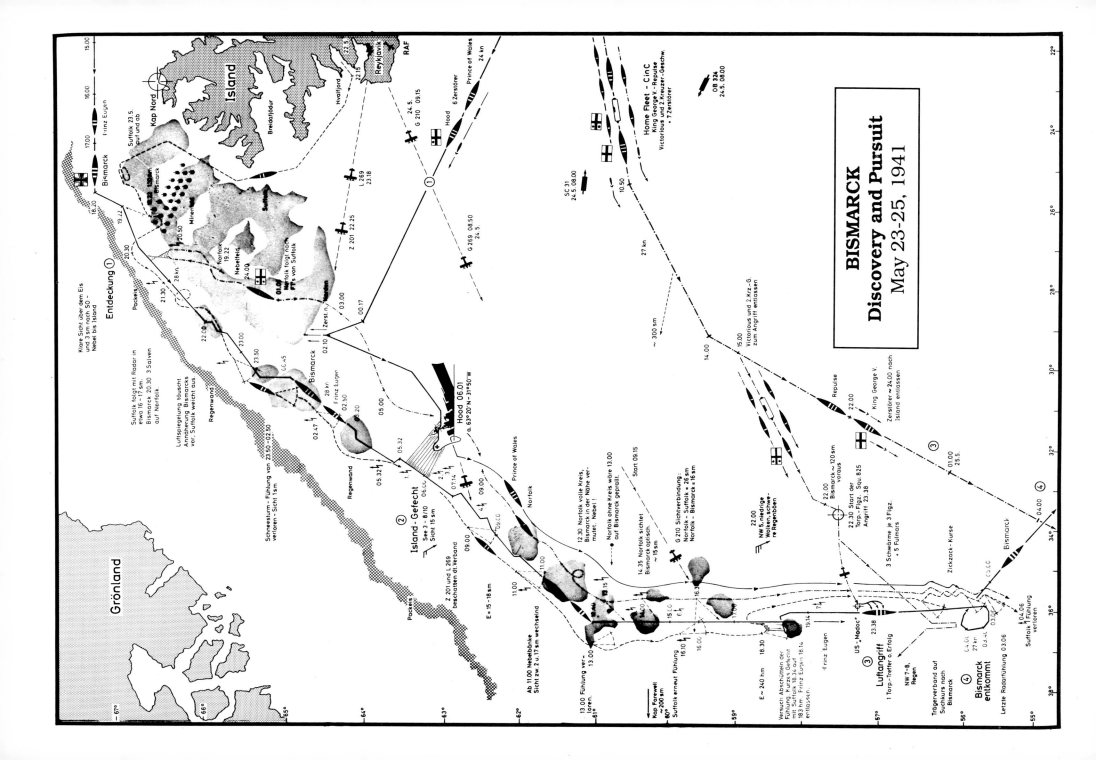

BISMARCK
Discovery and Pursuit
May 23-25, 1941

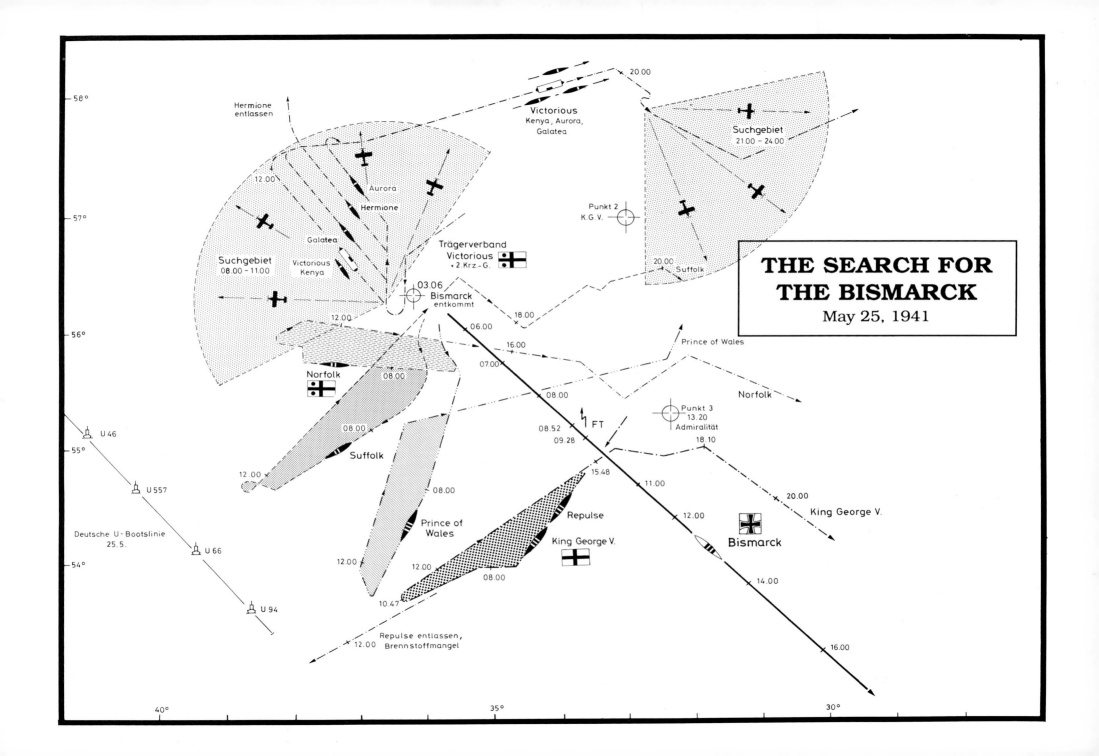

THE SEARCH FOR
THE BISMARCK
May 25, 1941

On Sunday, May 25, about 3:00 A.M., a few hours before the torpedo-plane attack, the BISMARCK was able to escape enemy surveillance. The search leader SUFFOLK lost radar contact with the German battleship. It almost seemed as if the BISMARCK could escape the concentrated encircling of the British naval forces. The German Fleet Chief Lütjens, unaware of the changed situation, sent off a long radio report just at this time, which could be tracked by the enemy with no trouble. Again luck was on the side of the German battleship, for the British could locate their opponent, the BISMARCK, only inexactly. Meanwhile it was 10:30 A.M. The British ships were running low on fuel, which seriously limited further pursuit. The battleship REPULSE had to be released to refuel. Chance often plays a decisive role in the history of peoples: the BISMARCK was discovered on May 26, at almost 10:30 A.M., exactly 31.5 (thirty-one and one-half) hours after contact had been lost, almost simultaneously by a "Catalina" flying boat and a carrier-based plane. But now she was so far ahead of the British units that only an attack by the planes of the aircraft carrier ARK ROYAL, which was within range, promised any chance of success. At 22,000 tons (with the same number of planes) the ARK ROYAL was only slightly smaller than the VICTORIOUS, while her armament of 16 11.4-cm guns, speed of 30 knots with 100,000 HP and crew of 1575 men were almost identical.

The takeoff for the decisive attack against the BISMARCK began about 7:00 P.M. on that May 26.

Between 8:30 and 9:15 P.M. the German battleship was attacked. Within this time, despite vigorous defensive fire, it received the fatal hit in the port rudder system, through which the ship lost its maneuverability. This hit and the reaction of the German fleet leadership have become the object of much speculation since then.

This picture shows one of the participating "Swordfishes" flying back to the ARK ROYAL after the attack.

None of the antiquated and slow planes was shot down in the attack.

The BISMARCK could no longer be held on course from this moment on. She moved in only barely controllable zigzag movements, on almost the opposite of the required course. Thus the British heavy units could close in quickly. The oil slick at the right side of the picture makes the BISMARCK's irregular course clear. This heating oil probably was still coming from the leak that the BISMARCK had sustained during the Iceland battle. Since then the BISMARCK had traveled more than 1500 nautical miles.

This photo was obviously taken from a reconnaissance plane at a time when the BISMARCK's final battle was just about to occur.

The Chronicle of the BISMARCK

from May 25 to 27 in outline form

Sunday, May 25, 1941 (Fleet Chief's Birthday)

12:28 A.M.:	BISMARCK reports attack by torpedo planes from aircraft carrier VICTORIOUS. Torpedo hit to starboard.
12:37 A.M.:	BISMARCK reports: Expect further attacks.
1:53 A.M.:	BISMARCK reports: Torpedo hit of no significance.
2:41 A.M.:	Group West reports: West U-boats have instructions to move eastward.
3:06 A.M.:	SUFFOLK's last radar contact: British contact lost for 31.5 (thirty one and one-half) hours . . .
7:00 A.M.:	BISMARCK reports: . . . one battleship, two heavy cruisers keeping contact again . . .
8:52-9:28 A.M.:	36-minute BISMARCK radio message: homed in on by British stations.
9:30 A.M.:	DORSETSHIRE leaves SL-74 convoy.
11:25 A.M.	Raeder radios birthday greetings to Admiral Lütjens.
ca. 11:45 A.M.:	Aboard BISMARCK: Lütjens speech plus that of the Commander (Captain Lindemann).
4:25 P.M.:	Hitler radios birthday greetings to Admiral Lütjens.
6:10 P.M.:	Admiral Tovey (Commander-in-Chief on British battleship KING GEORGE V) turns to southeast course.
11:44 P.M.:	Group West reports: Presumed continued course for west coast of France, even without enemy contact.

Monday, May 26, 1941

10:30 A.M.:	"Catalina" flying boat (Briggs) sights BISMARCK.
11:15 A.M.:	ARK ROYAL planes make contact—then SHEFFIELD too.
2:50-3:00 P.M.:	ARK ROYAL: Takeoff of 15 "Swordfish" torpedo planes. Mistaken attack on SHEFFIELD: Target confusion with BISMARCK!
5:47 P.M.:	SHEFFIELD sights BISMARCK.
7:03 P.M.:	BISMARCK reports: fuel situation urgent—when can I expect refueling?
7:48 P.M.:	U-556 reports: Battleship and ARK ROYAL in sight.
8:39 P.M.:	U-556 reports: Battleship and aircraft carrier on 115-degree course, high speed.
ca. 8:45 P.M.:	BISMARCK; Torpedo planes from ARK ROYAL in sight.
8:55-9:25 P.M.:	ARK ROYAL's "Swordfish" torpedo planes attack BISMARCK without losses . . .
9:03-9:05 P.M.:	BISKARCK hit aft (rudder system)—BISMARCK reports:

Unable to maneuver at ca. 47 degrees 40 minutes north-14 degrees 50 minutes west. Rudder jammed port-15.

9:15 P.M.:	BISMARCK reports: Square BE-6192—Torpedo hit aft!
10:38 P.M.:	Polish destroyer PIORUN sights BISMARCK and comes under fire.
11:15 P.M.:	BISMARCK changes course from southeast to northwest.
11:24 P.M.:	Captain Vian organizes his five destroyers (COSSACK, MAORI, PIORUN, SIKH, ZULU) for the planned (and also carried out) night attack on BISMARCK, but comes immediately under fire from BISMARCK.
11:40 P.M.:	BISMARCK reports: Ship unable to maneuver—we'll fight to the last shell—long live the Führer.
11:58 P.M.:	BISMARCK reports: to the Führer of the German Reich Adolf Hitler—we'll fight to the last in trust in you, our Führer, and in rock-hard trust in Germany's victory.

Tuesday, May 27, 1941

1:53 A.M.:	Hitler radios Lütjens, BISMARCK: I thank you in the name of all the German people—Adolf Hitler—to crew of battleship BISMARCK: all Germany is with you—whatever can still be done, will be done—your devotion to duty will strengthen our people in their fight for existence—Adolf Hitler.
2:21 A.M.:	BISMARCK reports: Recommend conferring of Knight's Cross on Corvette Captain Schneider for sinking of HOOD.
3:51 A.M.:	Radio message to BISMARCK A(rtillery) O(fficer) Corvette Captain Schneider: The Führer has conferred the Knight's Cross on you for the sinking of the Battleship HOOD—Heartiest good wishes—Commander of the Navy, Grand Admiral Raeder.
7:10 A.M.:	Last message from BISMARCK: Send U-boat to take war log book (KTB).
ca. 8:00 A.M.:	Alarm on BISMARCK, which is surrounded by the two British battleships KING GEORGE V (C-in-C Flagship) and RODNEY plus two cruisers DORSETSHIRE and NORFOLK—Weather: northwest wind (320 degrees) 6-7—sea: 4-5—Visibility up to 10 nautical miles.
8:15 A.M.:	NORFOLK sights BISMARCK.
8:49 A.M.:	Final battle begins: BISMARCK returns the British fire.
ca. 9:02 A.M.:	BISMARCK: Turrets A and B out of action.
ca. 9:12 A.M.:	BISMARCK: Forward command post out of action.
ca. 9:18 A.M.:	BISMARCK: After command post out of action.
ca. 10:00 A.M.:	All weapons silent on BISMARCK—the ship is a wreck.
10:36 A.M.:	BISMARCK founders at about 48 degrees 10 minutes north-16 degrees 12 minutes west. C-in-C (Admiral Tovey) reports: I should like to express my highest admiration for the thoroughly brave fight of the BISMARCK in a hopeless position . . . Rescuing survivors . . .

The BISMARCK's Last Battle

In the BISMARCK's final battle, the British battleship RODNEY opened fire on the BISMARCK at 8:47 A.M. One minute later the battleship KING GEORGE V followed. Another minute later the BISMARCK returned fire (at a distance of 220 hm), at first at the RODNEY. At first the BISMARCK defended itself energetically. In clouds of black smoke, the British battleship tries to avoid the BISMARCK's fire (see picture on page 152). The picture on the previous pages show the turret-high columns of water thrown up by heavy British shells behind the stern of the German battleship; finally, on this page the burning wreck is shown, obviously at a point in time when the heavy British units had already ceased fire.

After somewhat more than one hour, the BISMARCK could no longer fire, since all guns were out of action.

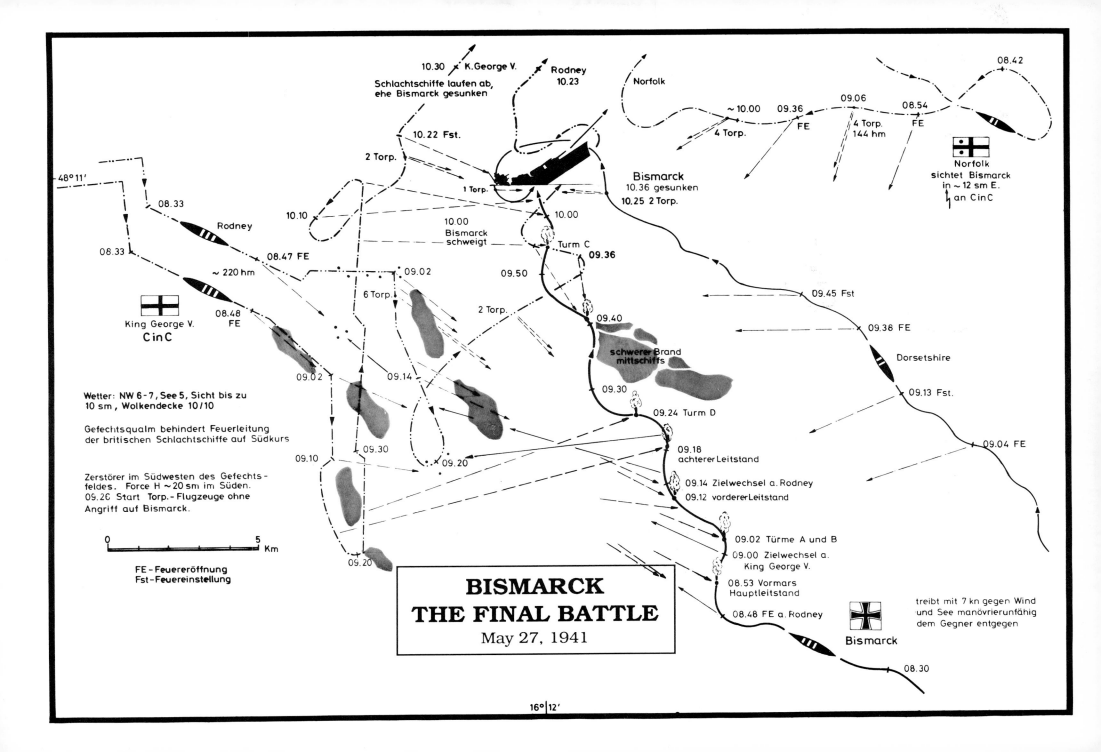

10.30 ✕ K.George V. Rodney 10.23
Schlachtschiffe laufen ab,
ehe Bismarck gesunken

Norfolk

08.42

10.22 Fst.

2 Torp.

~10.00 09.36 09.06 08.54
4 Torp. FE 4 Torp. FE
         144 hm

Bismarck
10.36 gesunken
10.25 2 Torp.

1 Torp.

Norfolk
sichtet Bismarck
in ~ 12 sm E.
an CinC

48° 11'

08.33

Rodney

10.10

10.00
Bismarck
schweigt

10.00

09.45 Fst

08.33

08.47 FE

Turm C 09.36

~ 220 hm

09.02

09.50

2 Torp.

09.38 FE

King George V.
CinC

08.48
FE

6 Torp.

09.40

09.30

Dorsetshire

09.13 Fst

schwerer Brand
mittschiffs

Wetter: NW 6-7, See 5, Sicht bis zu
10 sm, Wolkendecke 10/10

09.02

09.14

09.24 Turm D

Gefechtsqualm behindert Feuerleitung
der britischen Schlachtschiffe auf Südkurs

09.04 FE

09.18
achterer Leitstand

Zerstörer im Südwesten des Gefechts-
feldes. Force H ~20 sm im Süden.
09.20 Start Torp.-Flugzeuge ohne
Angriff auf Bismarck.

09.10

09.30

09.20

09.14 Zielwechsel a. Rodney
09.12 vorderer Leitstand

09.02 Türme A und B

09.00 Zielwechsel a.
King George V.

0   5
                        Km

08.53 Vormars
Hauptleitstand

08.48 FE a. Rodney

FE - Feuereröffnung
Fst - Feuereinstellung

09.20

BISMARCK
THE FINAL BATTLE
May 27, 1941

treibt mit 7 kn gegen Wind
und See manövrierunfähig
dem Gegner entgegen

Bismarck

08.30

16° 12'

The British heavy cruiser DORSETSHIRE—Displacement 9975 tons—Armament: 8 20.3-cm guns; 8 10.2-cm anti-aircraft guns, 4 4.7-cm anti-aircraft guns, 8 53.3-cm torpedo tubes, one airplane—Top speed: 32.2 knots—Launched: 1929—had the opportunity to fire three torpedoes at the BISMARCK about 10:30 A.M., but they could not sink the giant ship. The BISMARCK finally went down as a result of its own explosion in the turbine room on May 27, 1941 at 10:36 A.M. This assertion, though, cannot be made with complete certainty. The English ship then took part in the rescue action of the BISMARCK survivors, but had to break off this action then on account of submarine alarm after picking up 85 survivors.

In all, 111 crewmen of the BISMARCK were rescued.

Almost 2000 men died with the BISMARCK, most of them at a time when the ship was no more than a defenseless wreck. Until early in this century it was thoroughly customary and honorable for ship commanders to surrender their ships to the enemy when they were incapable of fighting. But the sea battles of World Wars I and II were obviously fought under different moral concepts, so that this alternative—and this is particularly clear in the case of the BISMARCK—could no longer be considered.

The destruction of the BISMARCK and the death of its crew have motivated many artists. The great German maritime painter Claus Bergen (1885-1964) probably had the best artistic relationship to this historic event. Through the intermediary influence of co-author Bodo Herzog, this oil painting has hung since 1963 in the Naval School in Flensburg-Mürwick. It is a gift of the now-retired salvage assessor Dr. Hermann Reuch.

Final Observations

The battleship BISMARCK was in service 277 days, and its only operation lasted no more than 215 hours. The German Navy possessed four major warships for only 92 days; in addition to the BISMARCK these were the battle cruisers SCHARNHORST and GNEISENAU and the battleship TIRPITZ. Thoughts of prestige and the fear that the German major warships of World War II could prove to be similarly superfluous to those of World War I led to Operation "Rheinübung." There is no comparable naval operation in which irreplaceable human lives and ships were sent into action similarly unprotected, independent of the fact that the German military forces could not offer this protection. Even if all the battleships had been ready to fight, any safety provision would have been insufficient to say the least, without going into the lack of air support. The deployment of aircraft carriers by the British had changed the situation of the German Navy considerably.

Grand Admiral Raeder, at that time the Commander of the Navy, has been criticized vigorously for that reason. Cajus Bekker writes in his book *Verdammte Sea* ("Damned Sea") as follows: "Battleships must strike! Certainly under such a lucky star as on May 24, 1941. The serious error of ending the fight with half success can be charged to the theoretical-utopian command given by the naval command, decisively formed by the Grand Admiral himself."

Cajus Bekker goes on to say: "Raeder again needs a diversionary effect in the Atlantic . . . It must have been learned with astonishment that the layman Hitler evaluates the deployment of the heavy ships more soberly and much more realistically than the specialists of the naval command."

When the pressure of later war experiences changed the German naval command's concept of sea strategy, they were not yet willing, after the loss of the BISMARCK and the buildup of a support system, to evaluate the realities soberly.

In this respect, an excerpt from a secret command item (G.Kdos. 24/3 (MDV No. 601)—Source: Federal Archives-Military Archives) from the year 1942, according to which such operations were still given priority, is of significance:

OPERATIVE FINAL OBSERVATION

The operative conclusions that must be drawn from the "Bismarck" operation are influenced by time, which means they must be brought into agreement with the overall war situation of the time, our own situation and that of the enemy.

For German conduct of the sea war with overwater forces, the most important theater of war in 1941 was the Atlantic Ocean. The battleship and heavy cruiser operations conducted fortunately and successfully against merchant shipping in the winter of 1940-41 strengthened the confidence of the sea-war leadership that continued deployment of overwater forces in the northern and central Atlantic Ocean would give the submarine war on merchant shipping worthwhile and essential support and force the enemy to a power-limiting division of his defensive forces. This effort caused the sea-war leadership to order the "Bismarck" operation, although the battleships "Gneisenau" and "Scharnhorst" were not at that time, and the battleship "Tirpitz" was not yet, ready for action, and although the short nights of the summer months had to make an unnoticed breakthrough of the "Bismarck" battle group into the area of operations difficult, a disadvantage that never-

theless, in the opinion of the sea-war leadership, was in part nullified, that—as it in fact occurred in the "Bismarck" operation—during the spring months bad-visibility weather is often encountered during the spring months. The sea-war leadership was well aware that, by the very nature of conducting this type of war, minor causes could bring about major results at any time, and that despite all care on the part of land and sea leadership, the fortunes of war could always change.

The basic idea of waging war with overwater fighting forces involves surprise and constant changes of operational areas; it is thus, even when it is conducted by battleships, a true cruiser warfare, in which fighting against enemies of equal power must always be merely a means and a goal. The main task of the navy is and remains the interruption of shipping to England. All fighting forces suited to this purpose must take part in this task. Only in their mutual completion and reciprocal effect is there any release for the other. The fighting of our U-boats also requires the deployment of overwater fighting forces appropriate to their potential, for the native battle group predominantly in the North Atlantic. The enemy fears this type of warfare particularly, since it is capable of bringing disorganization to his convoy system hitherto in use and thus bring about further major and unbearable disturbances in his supply traffic.

For future operations, a series of lessons derive from the "Bismarck" operation:

1. For the conduct of the operation, greater emphasis must be placed on the moment of surprise, in order to prevent the enemy from learning of the advance and make the advance of his forces difficult. Therefore no more advances from the German home ports with the possibility of sighting and reporting by agents, even from the Belt and the Sund, but transfer of the ships to Drontheim weeks before the beginning of the operation, and breakthrough during suitable weather conditions, because the breakthrough from a harbor on the Atlantic coast is more surprising.

2. Despite the unfortunate results of the "Bismarck" operation, the small, mobile battle group, easier to supply, even a single battleship or cruiser, is the most suitable unit for this kind of cruiser warfare. Its goal is not to go into battle, but rather to intervene in the enemy's sea connections unnoticed.

When possible, the battle group must consist of homogeneous units in respect to firepower, speed and sea range.

3. The difficulty of refueling caused by enemy aircraft carriers and radar equipment requires a different positioning of our tankers. For one thing, these ships must not remain at the supply points, but must be moved to very faraway areas when possible. For another, the provision of a fast supply ship is required, which thanks to its high cruising speed is capable of crossing large distances quickly in cases of fuel shortage.

4. Midsummer, with its short light nights, is particularly unfavorable for operations in the northern area. Its disadvantages are not mitigated by the possibility of meeting fog frequently in the far north at this time of year.

Under the slogan "Protect Prestige", the fate of the BISMARCK was enlighteningly portrayed in a candid defense, which led in part to the rise of a mythology. On that subject, Prof. Dr. Michael Salewski writes in his book "Die deutsche Seekriegsleitung" (The German Sea-War Command):

"The fight and loss of the BISMARCK rank among the best-known, most controversial, most frequently described events in the history of naval warfare in World War II. An endless mass of foreign and German literature has taken up the subject; scientific, half-scientific, popular and sensationalistic investigations, 'factual accounts' and films have pursued every detail of the matter . . . everything has been investigated to the last detail . . . it all resembles a play whose drama is gripping and moving . . . The loss of the BISMARCK was . . . nothing extraordinary, no 'stroke of fate' . . ."

Photo Credits and Bibliography

For their friendly help and support, as well as making pictorial material available, the authors thank:

Frigate Captain (retired) Paul Schmalenbach, Altenholz; Foto-Kino Paul Binder, Oberhausen, Rheinland; Blohm & Voss A.G., Public Relations Department, Hamburg 1; Imperial War Museum, London; Federal Archives, Koblenz; Ulf Busch, Kiel; Foto-Drüppel, Wilhelmshaven; Profile Publications Ltd., Berkshire, England; Ferdinand Urbahns, Eutin.

Marine Cartographic Engineer Helmut Fechter, Osterode, Harz, provided the maps based on the present state of research.

Heinrich Schmitt, of the Federal Archives, Koblenz: see page 119.

Photo captions of pictures on pages 114 right, 115, 117, 120 right, 122 and 123 (PK-Lagemann) by Fritz-Otto Busch, "PRINZ EUGEN in its first battle"—Gütersloh 1943.

Translator's note:

Just before I started to translate this book, the October 1989 issue of "National Geographic" appeared, including the article *Finding the Bismarck*. This article describes exploration that culminated in the discovery of the wreck of the BISMARCK, includes actual photographs of portions of the wreck, a chart of the final battle, and other details such as a photo of the only officer to survive the sinking, who is still alive, and the statement that 115 of the ship's crew, not 111 as stated in this book, were rescued by the British. Readers will probably find this article of much interest.

Bibliography

P. Schmalenbach, *Die Geschichte der deutschen Schiffsartillerie*, Herford 1968.

W. Hadeler, *Kriegsschiffsbau*, Darmstadt 1968; 2 volumes of C.Bekker (H. D. Berenbrock), *Verdammte See*, Oldenburg/Hamburg 1971.

E. Bradford, *The Mighty Hood*, London/New York/Sydney/Toronto 1974.

J. Brennecke, *Schlachtschiff BISMARCK*, Herford 1960, 3rd ed.

F. O. Busch, *PRINZ EUGEN im ersten Gefecht*, Gütersloh 1943.

R. G. Robertson, *HMS/Battle Cruiser 1916-1941*, (Warship Profile No. 19).

P. Schmalenbach, *Kriegsmarine BISMARCK*, (Warship Profile No.18).

R. Grenfell, *Jagd auf die BISMARCK*, Tübingen 1958.

Bodo Herzog, *Die deutsche Kriegsmarine im Kampf*, Dorheim 1969.

Ploetz, *Geschichte des Zweiten Weltkrieges*, Würzburg 1960.

S. Breyer, *Schlachtschiffe und Schlachtkreuzer 1905-1970*, Munich 1970.

E. Gröner, *Die deutschen Kriegsschiffe 1815-1945*, Munich 1966.

Rheinmetall GMBH, *Taschenbuch für den Artilleristen*, Düsseldorf 1961.

Rheinmetall GMBH, *Waffentechnisches Taschenbuch"*, Düsseldorf 1973.

H. H. Adams & S. Rohwer, *Uberwasser-Operationen im Atlantik* . . .

F. O. Busch, *Das Geheimnis der BISMARCK*, Hannover 1950.

G. Hummelchen, *Handelsstörer* . . ., Munich 1967, 2nd ed.

P. K. Kemp, *The Case of the BISMARCK* . . ., (Historical of the Second World War, No. 5, 1967).

S. W. Roskill, *The War at Sea 1939-1945, Vol. I: The Defensive* . . ., London 1961, 5th ed.

M. Salewski, *Die deutsche Seekriegsleitung 1935-1945* . . ., Vol. I-III, Frankfurt am Main 1970/73/75.

K. Assmann, *Deutsche Schicksalsjahre* . . ., Wiesbaden 1950.

H. Burkhardt, *Die Entwicklung des Schiffbaumaterials der Deutschen Kriegsmarine*, (Marine-Rundschau No. 2, 1961).

K. F. Ludwig, *Sie bauten Deutschlands Flotte auf*, (KöhlersFlotten-Kalender 1962).

K. Stange, *Die Flottenverträge von 1922 bis 1933* . . ., (Marine-Rundschau No. 3, 1968).

G. Thomas & W. Birkenfeld, *Geschichte der deutschen Wehr-und Rüstungswirtschaft* . . ., Boppard am Rhein 1966.

K. H. Ludwig, *Technik und Ingenieur im Dritten Reich*, Düsseldorf 1974.

W. Hadeler, *Kriegsschiffbau*, Darmstadt 1968, 2 volumes.

Press reports (selected) on the BISMARCK—from launching to Operation "Rheinübung": *"Völkischer Beobachter"*, *"Rheinisch-Westfälische Zeitung"*, *"Kölnische Zeitung"*, *"Frankfurter Zeitung"*, *"The Times"*, *"Neue Zürcher Zeitung"*, *"Das Reich"*, *"Der Angriff"*, *"The Illustrated London News"*, *"National-Zeitung"*, *"Die Kriegsmarine"*, *"Marine-Rundschau"*, *"West-Front"*, *"Weser-Zeitung mit Bremer Nachrichten"*, *"Hannoverscher Anzeiger."*